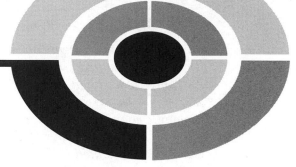

ADVANCED STATISTICS
DEMYSTIFIED

Demystified Series

ADVANCED STATISTICS
DEMYSTIFIED

LARRY J. STEPHENS

McGRAW-HILL
New York   Chicago   San Francisco   Lisbon   London
Madrid   Mexico City   Milan   New Delhi   San Juan
Seoul   Singapore   Sydney   Toronto

The **McGraw·Hill** Companies

Cataloging-in-Publication Data is on file with the Library of Congress

Copyright © 2004 by The McGraw-Hill Companies, Inc. All rights reserved. Printed in the United States of America. Except as permitted under the United States Copyright Act of 1976, no part of this publication may be reproduced or distributed in any form or by any means, or stored in a data base or retrieval system, without the prior permission of the publisher.

4 5 6 7 8 9 0   DOC/DOC   0 1 0 9 8 7

ISBN 0-07-143242-6

The sponsoring editor for this book was Judy Bass and the production supervisor was Pamela A. Pelton. It was set in Times Roman by Keyword Publishing Services Ltd. The art director for the cover was Margaret Webster-Shapiro; the cover designer was Handel Low.

Printed and bound by RR Donnelley.

Minitab is a registered trademark of Minitab Inc.

To my Mother and Father, Rosie and Johnie Stephens

Larry J. Stephens is a professor of mathematics at the University of Nebraska at Omaha. He has over 25 years of experience teaching mathematics and statistics. He has taught at the University of Arizona, Gonzaga University, and Oklahoma State University, and has worked for NASA, Livermore Laboratory, and Los Alamos National Laboratory. Dr. Stephens is the author of *Schaum's Outline of Beginning Statistics* and co-author of *Schaum's Outline of Statistics*, both published by McGraw-Hill. He currently teaches courses in statistical methodology, mathematical statistics, algebra, and trigonometry.

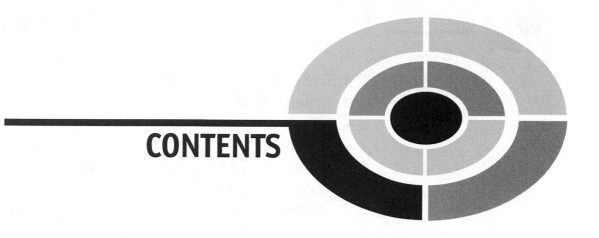

CONTENTS

CONTENTS

x

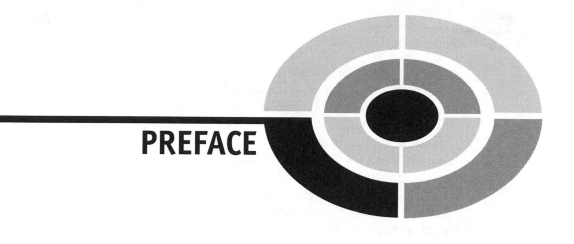

PREFACE

Since receiving my Ph.D. in Statistics from Oklahoma State University in 1972, I have observed unbelievable changes in the discipline of statistics over the past 30 years. This change has been brought about by computer/statistical software. With the introduction of Minitab in 1972, a tremendous change (for the better) has occurred in statistics. I wish to thank Minitab for permission to include output from the company's software in this book. (MINITAB^{TM} is a trademark of Minitab, Inc., in the United States and other countries and is used herein with the owner's permission. Minitab may be reached at www.Minitab.com or at the following address: Minitab Inc., 3081 Enterprise Drive, State College, PA 16801-3008.) The output, including dialog boxes and various pull-down menus included in the book, is taken from Release 14. Pull-down menus are indicated by the symbol $\Rightarrow$. Snapshots of dialog boxes are included throughout the text. The reason that Minitab is chosen is that it is widely available at colleges and universities. It is also widely available at reasonable prices for students to install on their home computers.

Output from Microsoft Excel is also included in the book. This software has been available since 1985, and is now widely available on home computers. In surveys conducted in my classes, I find that nearly all my students have Excel on their home computer. Excel has many built-in statistical routines even though it is not primarily statistical software. If a statistical procedure is not built-in, the worksheet will make the computations for the procedure easy to carry out. I will often illustrate a statistical procedure for both Minitab and Excel. Some readers of the text will have only one of these systems available.

There are many areas of application for statistics. I have tried to give examples and exercises that are realistic. Many are taken from recent issues of *USA Today*. Not only do medical, engineering, and business areas of applications exist but most research areas have statistical applications. Most graduate programs that involve a thesis project require some knowledge of statistics. Any study that involves the gathering of data and inferences from that data requires knowledge of statistics. People from disciplines ranging from art to zoology require statistical analysis, and anything that makes that analysis easier is welcome. The data given in the problems has been chosen to be as realistic as possible. The purpose of the examples is to illustrate how to use statistics to make conclusions, but the reader is reminded that, if the experiment were actually conducted, results different from those stated might have occurred.

The background required to understand this book is a good command of high school algebra and good computer skills. It is also helpful if the reader has had an introduction to a statistics course at the high school or college level. The author welcomes comments concerning the book at Lstephens@Unomaha.edu.

I wish to thank my wife, Lana, for her helpful discussions of problems and concepts. She has had a year of graduate level statistics and understands the difficulties of statistics. Her help is indispensable. Thanks also goes to my friend and computer consultant, Stanley Wileman. I have never had a computer-related question he could not answer. Thanks also goes to Richard Cook, Editorial Production Manager, and Ian Guy, copy editor, Keyword Publishing Services, United Kingdom. And finally, thanks to Judy Bass, Senior Acquisitions Editor, and her staff at McGraw-Hill.

Larry J. Stephens

Introduction: A Review of Inferences Based on a Single Sample

I-1 Large Sample ($n > 30$) Inferences About a Single Mean

An introductory statistics course usually covers an introduction to estimation and tests of hypotheses about one parameter based on one sample. This is the background that the reader is assumed to have. This chapter will give a review of the introduction to inference about a single parameter. We shall review inferences about population mean, μ (large n), μ (small n), population proportion, p, and population standard deviation, σ.

> The five books contained in the bibliography contain additional discussion of the review material contained in this chapter as well as material relating to the integration of Excel and MINITAB with statistical concepts.

"Large sample," when making inferences about a single population mean, is taken to mean that the sample size $n > 30$. The sample mean, $\bar{x}$, is a *point estimate* or a single numerical estimate for the population mean, μ. The *interval estimate* is a better estimate because it gives the reliability associated with the estimate. A $(1 - \alpha)$ large sample confidence interval estimate for μ is $(\bar{x} \pm z_{\alpha/2}(s/\sqrt{n}))$.

A hypothesis test for μ is of the following form: the null hypothesis H_0: $\mu = \mu_0$ versus one of the three research hypotheses: H_a: $\mu < \mu_0$, $\mu > \mu_0$, or $\mu \neq \mu_0$. The test statistic is $Z = (\bar{x} - \mu_0)/(s/\sqrt{n})$. The statistic, Z, has a standard normal distribution.

The statistical test is performed by one of three methods called the *classical method*, the *p-value method*, or the *confidence interval method*.

EXAMPLE I-1
We will begin with an introductory example and review terms in the context of the example. An article entitled "Study: Candles causing more home fires than ever before" recently appeared in *USA Today*. The article estimated that sales of candles and candle accessories reached an estimated high of \$2.3 billion annually in the late 1990s. Suppose a study was conducted and the amount spent in dollars per person for the past year on candles and accessories for a sample of 300 people across the United States was collected. The results of the study are given in Table I-1. Use the data to set a 95% confidence interval on μ, the mean amount spent on candles for the American population.

SOLUTION
The 300 data points were entered into column C1 of the Minitab worksheet as shown in Fig. I-1.

The Minitab pull-down **Stat $\Rightarrow$ Basic Statistics $\Rightarrow$ Display Descriptive Statistics** gave the dialog box shown in Fig. I-2. If Statistics is selected in this dialog box, Fig. I-3 gives another dialog box and the many choices available. The choices Mean, Standard deviation, SE Mean, and N are selected.

Introduction

Table I-1 Annual spending for candles and candle products for each person in a sample of 300.

15	7	29	23	29	3	11	18	14	24	14	16	28	3	16
7	29	24	7	20	25	17	19	28	29	29	27	12	1	0
5	12	3	2	13	20	15	23	8	7	4	21	25	19	28
1	16	8	16	18	20	0	19	17	16	11	13	26	30	26
1	25	5	19	19	0	26	6	11	18	6	3	30	20	20
10	26	27	2	29	11	0	11	0	25	22	8	30	23	18
0	27	29	21	11	10	22	23	23	9	29	15	19	9	18
17	16	16	15	28	14	10	22	29	20	30	0	27	8	20
27	22	5	5	15	30	19	10	27	16	27	10	27	12	25
22	17	10	13	17	7	11	18	18	28	7	27	18	10	24
13	11	9	6	13	18	17	14	28	23	26	25	6	11	1
13	0	26	29	11	27	3	28	10	23	11	20	17	18	10
10	7	24	3	27	3	8	19	8	4	17	18	2	14	28
29	11	19	24	3	13	9	30	12	8	25	2	27	10	30
5	1	7	4	28	5	24	26	16	21	25	3	28	22	18
23	14	14	28	1	16	13	18	22	27	29	14	20	0	16
11	21	2	2	23	16	24	16	6	24	1	0	21	2	18
27	10	6	23	25	22	7	24	29	17	4	9	6	21	8
7	27	6	28	28	7	8	2	23	1	10	26	22	5	27
30	18	12	7	15	11	24	1	30	23	16	19	14	24	22

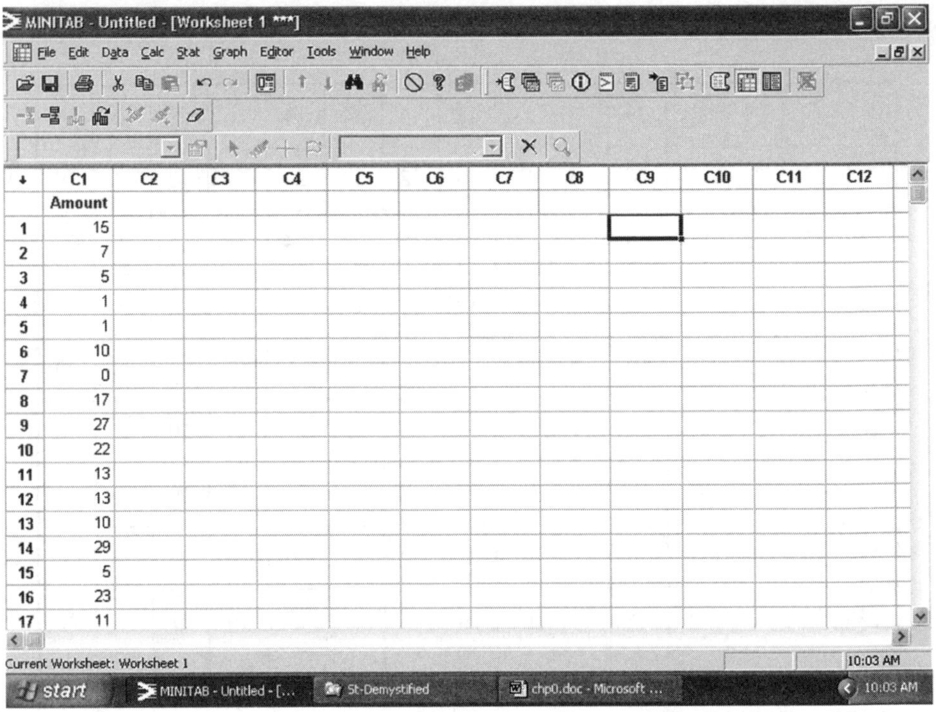
MINITAB - Untitled - [Worksheet 1 ***]
File Edit Data Calc Stat Graph Editor Tools Window Help
C1 C2 C3 C4 C5 C6 C7 C8 C9 C10 C11 C12
Amount
1 15
2 7
3 5
4 1
5 1
6 10
7 0
8 17
9 27
10 22
11 13
12 13
13 10
14 29
15 5
16 23
17 11
Current Worksheet: Worksheet 1
10:03 AM
start
MINITAB - Untitled - [...
St-Demystified
chp0.doc - Microsoft ...
10:03 AM

Fig. I-1.

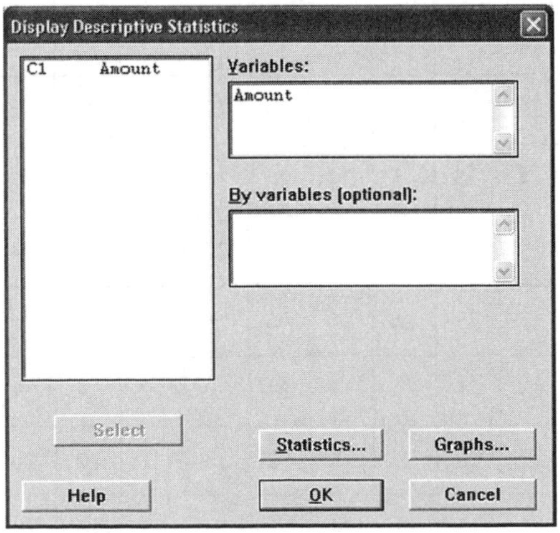
Display Descriptive Statistics
C1 Amount
Variables:
Amount
By variables (optional):
Select
Statistics...
Graphs...
Help
OK
Cancel

Fig. I-2.

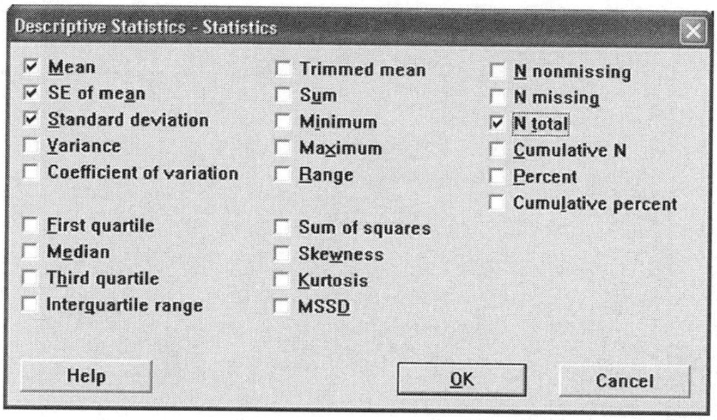

Fig. I-3.

The output below is produced.

Descriptive Statistics

Variable	Total Count	Mean	SE Mean	StDev
C1	300	15.983	0.518	8.979

The sample mean, $\bar{x} = \$15.98$, is called a *point estimate* for the population mean μ. In other words, $\$15.98$ was the average amount spent per person for the sample of 300. The symbol μ represents the mean for the United States population and the symbol $\bar{x}$ represents the mean of the sample. A 95% *confidence interval estimate* for the population mean is obtained by the Minitab pull-down **Stat $\Rightarrow$ Basic Statistics $\Rightarrow$ 1-sample Z**. The dialog box is filled out as shown in Fig. I-4.

The following output is obtained as a result of clicking OK in Fig. I-4.

One-Sample Z: Amount
The assumed standard deviation = 8.979

Variable	N	Mean	StDev	SE Mean	95% CI
Amount	300	15.9833	8.9788	0.5184	(14.9673, 16.9994)

The interval ($14.97, $17.00) is a 95% confidence interval for μ.

EXAMPLE I-2
Rather than estimate the mean amount spent per year on candles, it is possible to test the hypothesis that the mean amount spent equals a particular amount. Suppose the null hypothesis is that the mean amount spent annually

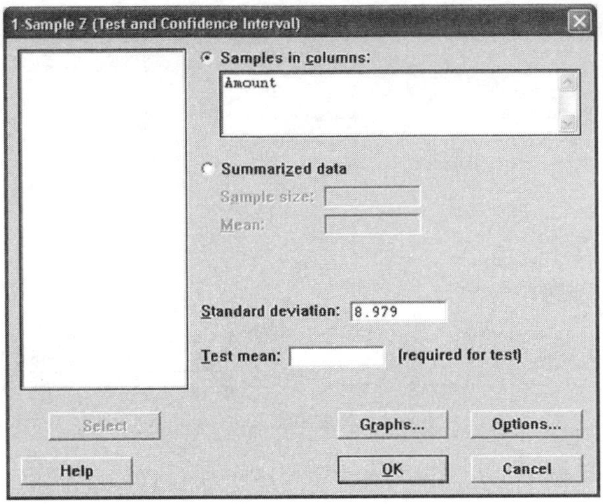

Fig. I-4.

per person is $20 versus it is not $20. The null and alternative hypotheses are stated as

$$H_0: \mu = \$20 \quad \text{and} \quad H_a: \mu \neq \$20$$

SOLUTION
The same sample data is used to perform the test of the hypothesis. Rather than estimate the mean, a hypothesis is tested concerning the mean amount spent on candles and candle accessories. Assuming that the null hypothesis is true, the *central limit theorem* assures us that $\bar{x}$ has a normal distribution with mean $\mu_0 = \$20$. The standard deviation of $\bar{x}$, called the *standard error of the mean*, is equal to $\sigma/\sqrt{n}$. If the population standard deviation σ is unknown, it is estimated by the sample standard deviation ($S = 8.9788$) and the *estimated standard error of the mean* is $S/\sqrt{n} = 8.9788/\sqrt{300} = 0.5184$. The above printout gives the estimated standard error. The *test statistic* is $Z = (\bar{x} - \mu_0)/(S/\sqrt{n})$. For samples larger than 30, the test statistic has a standard normal distribution. The *computed test statistic* is equal to $(15.9833 - 20)/0.5184 = -7.75$. Assuming a *level of significance* $\alpha = 0.05$, the *rejection region* is $|Z| > 1.96$ and the null hypothesis would be rejected since $Z = -7.75$ falls within the rejection region. The values -1.96 and 1.96 are called the *critical values*. They separate the rejection and non-rejection regions from each other. The conclusion is that the population mean is less than $20 because the computed test statistic falls in the rejection region on the low side of μ.

This method of testing hypotheses is called the *classical method* of testing hypotheses. The critical values are found and they separate the rejection and non-rejection regions. If the computed test statistic falls in the rejection region, then the null hypothesis is rejected.

> The logic is that the test statistic is not very likely to fall in the rejection region when the null hypothesis is true. The probability of falling in the rejection region when the null hypothesis is true is α. If the computed test statistic falls in the rejection region, reject the null hypothesis and accept the research hypothesis as true. Many students learn the mechanics of statistics, but do not comprehend the logic involved.

EXAMPLE I-3

Another method of testing hypotheses is the *confidence interval method*. To test H_0: $\mu = \$20$ and H_a: $\mu \neq \$20$ with $\alpha = 0.05$, form a confidence interval of size $1 - \alpha = 0.95$. If the interval does not contain the value $\mu_0 = \$20$, then reject H_0 at level of significance $\alpha = 0.05$.

SOLUTION

The 95% confidence interval is ($14.97, $17.00) and it does not contain $20. Therefore we reject the null. The same decision will be reached using either the classical method or the confidence interval method.

EXAMPLE I-4

A third method of testing hypotheses is the *p-value method*. This method computes the probability of observing a value of the test statistic that is at least as contradictory to the null hypothesis as the one computed using the sample data.

SOLUTION

In the present example, the *p*-value $= 2$ times $P(Z < -7.75)$. This probability is approximately 0. If the *p*-value $< \alpha$, then reject the null hypothesis. (In the case of a one-tailed research hypothesis, the probability is not doubled.) Note that the same decision is reached regardless of which of the three methods is used.

The standard normal curve (which has mean 0 and standard deviation 1 and is represented by the letter Z) is shown in Fig. I-5. The test statistic for large samples ($n > 30$), $Z = (\bar{x} - \mu_0)/(S/\sqrt{n})$, has a standard normal distribution. Recall that approximately 68% of the area under the standard

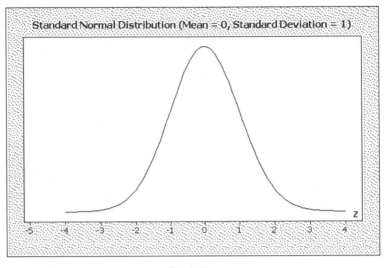

Fig. I-5.

normal curve is between -1 and $+1$, 95% of the area under the standard normal curve is between -2 and $+2$, and 99.7% of the area under the standard normal curve is between -3 and $+3$.

In the present example, the calculated test statistic is -7.75. Such a Z value is extremely unlikely because it is 7.75 standard deviations on the negative side of the Z distribution. Assuming the sample is random and representative, the reason for this unlikely value of Z is that the null hypothesis is likely to be false. The hypothesis (H_0: $\mu = \$20$) would be rejected in favor of (H_a: $\mu \neq \$20$).

EXAMPLE I-5
Suppose the hypothesis had been H_0: $\mu = \$15$ and H_a: $\mu \neq \$15$ with $\alpha = 0.05$.

SOLUTION
The computed test statistic would have been $(15.9833 - 15)/0.5184 = 1.90$. In this case the null hypothesis would not be rejected since the computed test statistic does not fall in the critical region. Alternatively, if the p-value is computed, it would equal 2 times $P(Z > 1.90) = 0.0574$ which is greater than 0.05. Remember, we only reject if the p-value $< \alpha$.

EXAMPLE I-6
Find the solution to the above problem using Excel.

SOLUTION

The Excel solution is shown in Fig. I-6. The 300 data values are entered into cells A1:A300. The classical as well as the *p*-value method for testing hypothesis is shown. The first third of the worksheet shows the computation of the test statistic. The second third shows the calculation of the critical values. The lower part of the worksheet shows the computation of the *p*-value.

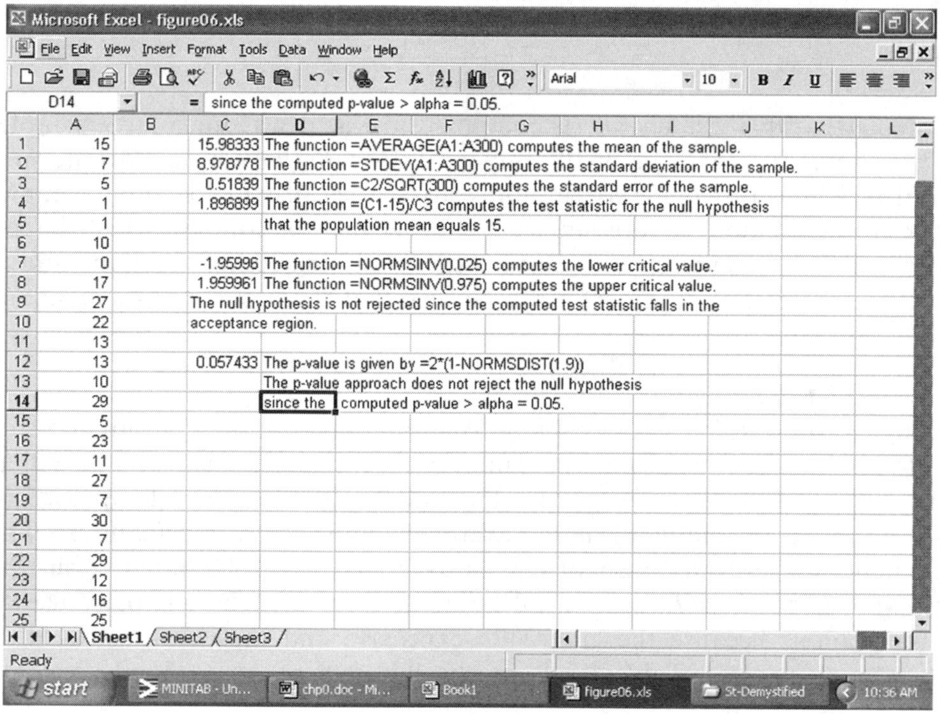

Fig. I-6.

I-2 Small Sample Inferences About a Single Mean

The test about a single mean for a small sample is based on a test statistic that has a student *t* distribution with degrees of freedom equal to $n-1$, where n is the sample size. It is assumed that the sample is taken from a normally distributed population. This assumption needs to be checked. If the population has a skew or is bi- or trimodal, for example, a non-parametric test should be used rather than the *t*-test. To estimate the population mean for small samples, use the following confidence interval:

$$\bar{x} \pm t_{\alpha/2} \frac{s}{\sqrt{n}}$$

To test the null hypothesis that the population mean equals some value μ_0, calculate the following test statistic:

$$T = \frac{\bar{x} - \mu_0}{s/\sqrt{n}}$$

Note that this test statistic is computed just as it was for a large sample. The difference is that, for small samples, it does not have a standard normal distribution (Z). It tends to be more variable because of the small sample. The T distribution is very similar to the Z distribution. It is bell-shaped and centers at zero. If a T curve with $n-1$ degrees of freedom and a Z curve are plotted together, it can be seen that the T curve has a standard deviation larger than one.

EXAMPLE I-7

Long Lasting Lighting Company has developed a new headlamp for automobiles. The high intensity lamp is expensive but has a lifetime that the company claims to be greater than that of the standard lamp used in automobiles. The standard has an average lifetime equal to 2500 hours. The company wishes to use a small sample to test H_0: $\mu = 2500$ hours versus the research hypothesis H_a: $\mu > 2500$ hours at $\alpha = 0.05$. The company uses a small sample because the lamps are expensive and they are destroyed in the testing process. Table I-2 gives the lifetimes of 15 randomly selected

Table I-2   Lifetimes of high intensity auto lamps.

3150	2669	2860
3033	2364	2423
2862	2575	2843
2827	3161	3134
3124	2570	2959

lamps of the new type. The lifetimes are determined by using the lamps until they expire.

SOLUTION

The data are entered into column C1 of the Minitab worksheet. The pull-down **Stat ⇒ Basic Statistics ⇒ 1-sample t** is used in Minitab to perform a 1-sample t test. Figure I-7 shows the dialog box.

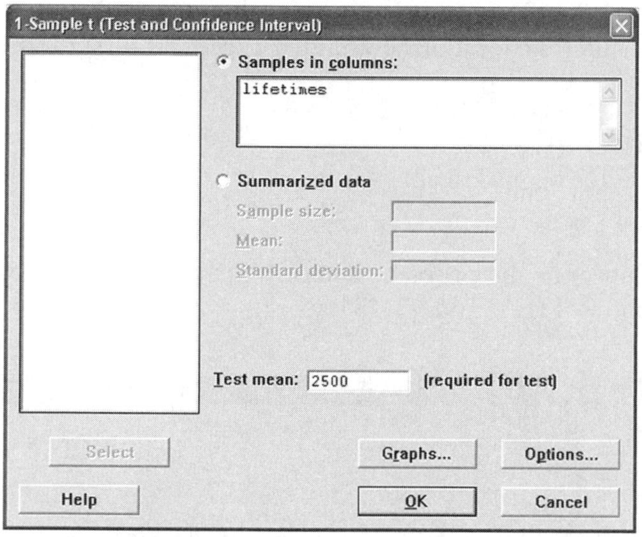

Fig. I-7.

The options dialog box is completed as shown in Fig. I-8. The alternative chosen is "greater than." It corresponds to H_a: $\mu > 2500$.

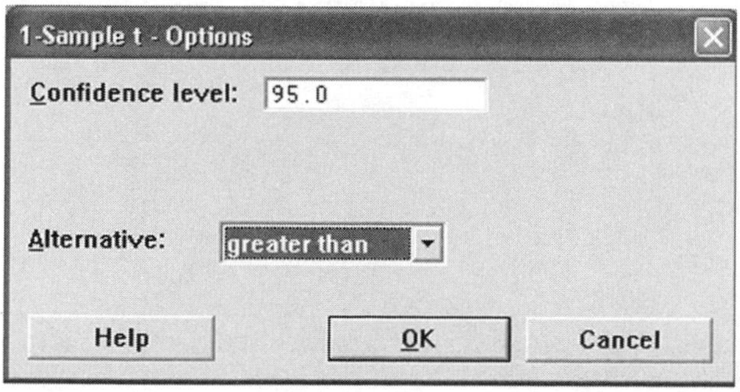

Fig. I-8.

The output is:

One-Sample T: lifetimes

Test of mu $= 2500$ vs > 2500

Variable	N	Mean	StDev	SE Mean	95% Lower Bound	T	P
lifetimes	15	2836.93	266.11	68.71	2715.92	4.90	0.000

The p-value, 0.000, is much smaller than $\alpha = 0.05$. It is concluded that the mean lifetime of the new lamps exceeds the standard lifetime of 2500 hours.

EXAMPLE I-8

Solve Example I-7 using Excel.

SOLUTION

The Excel solution to the problem is shown in Fig. I-9. The figure illustrates how the functions TINV and TDIST work.

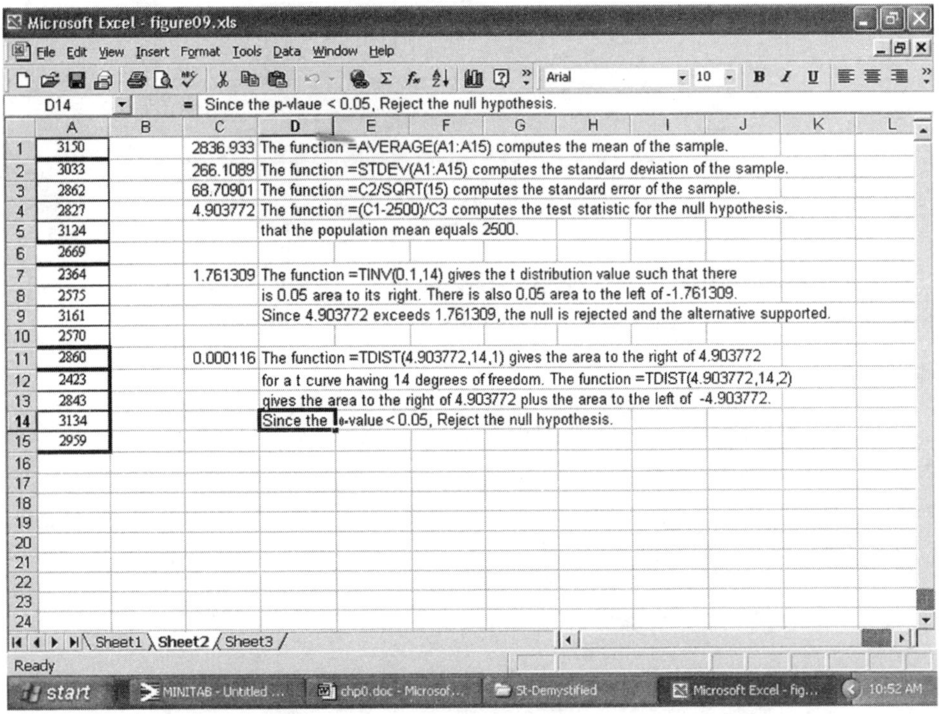

Fig. I-9.

I-3 Large Sample Inferences About a Single Population Proportion

The test statistic $Z = (\hat{p} - p_0)/\sqrt{p_0 q_0/n}$ is used to test the null hypothesis H_0: $p = p_0$ against any one of the alternatives $p < p_0$, $p > p_0$, or $p \neq p_0$. The hypothesized percent in the population with the characteristic of interest is p_0 and the percent in the sample of size n having the characteristic is $\hat{p}$. The normal approximation to the binomial distribution is used and is valid if $np_0 > 5$ and $nq_0 > 5$. If estimation is performed, the confidence interval is of the form $\hat{p} \pm z_{\alpha/2}\sqrt{\hat{p}\hat{q}/n}$. For the confidence interval to be valid, the sample size must be large enough so that $n\hat{p} > 5$ and $n\hat{q} > 5$. Almost all surveys meet this sample size requirement so that the standard normal approximation described above is valid in most real-world cases.

EXAMPLE I-9

A recent article in *USA Today* commented on the early eating habits of children. The percentages of 19- to 24-month old children who consumed the following foods at least once a day were: hot dogs, 25%; sweetened beverages, 23%; French fries, 21%; pizza, 11%; and candy, 10%. In order to test the 25% figure for hot dogs, a national telephone survey of 350 is taken and one of the questions is "Does your child consume a hot dog once a day?" 100 answered yes. The sample percent is $\hat{p} = 100/350 = 28.6\%$. We wish to test H_0: $p = 25\%$ versus H_a: $p \neq 25\%$ at $\alpha = 0.05$. Use the confidence interval method, the classical method, and the p-value method to perform the test.

SOLUTION

The pull-down **Stat $\Rightarrow$ Basic Statistics $\Rightarrow$ 1-proportion** gives the dialog box shown in Fig. I-10. This dialog box gives the sample size, 350, and the number who answered Yes, 100. In the Options portion of Fig. I-10, the confidence level is entered as 95%, the test proportion as 0.25, the research hypothesis as two-tailed (not =), and we check the box that indicates that the normal approximation to the binomial is to be used.

   The output created by Fig. I-10 is

```
Test of p=0.25 vs p not=0.25

Sample   X     N     Sample p    95% CI        Z-Value   P-Value
1        100   350   0.285714    (0.238387,    1.54      0.123
                                  0.333042)
```

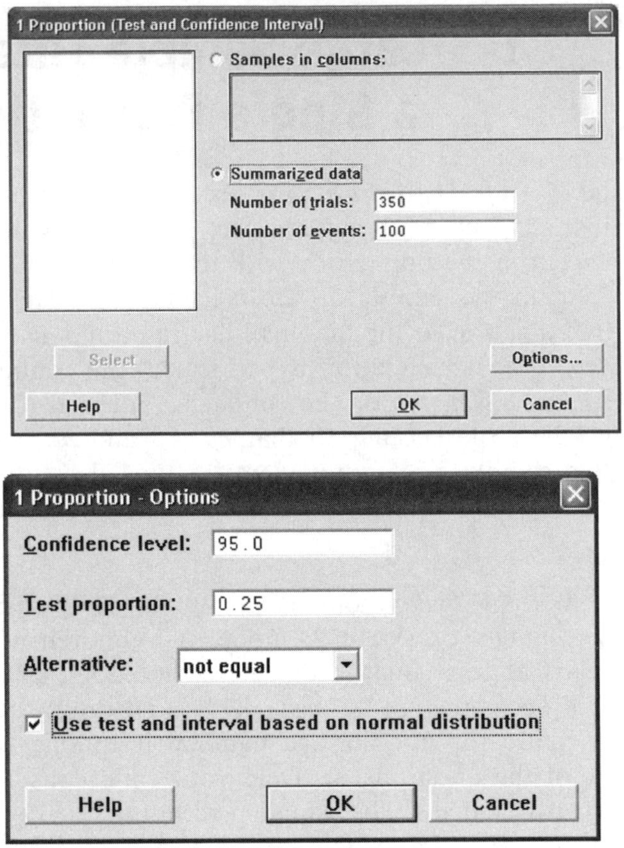

Fig. I-10.

A 95% confidence interval for p is $(0.238, 0.333)$. The computed test statistic is $Z = 1.54$ and the two tailed p-value is 0.123.

Using the three methods of testing, we find:

1. *Confidence interval method:* Since the 95% confidence interval for p (23.8%, 33.3%) contains $p_0 = 25\%$, you are unable to reject the null hypothesis at $\alpha = 0.05$.
2. *Classical method:* The rejection region is $Z < -1.96$ or $Z > 1.96$. The computed test statistic is 1.54, does not fall in the rejection region, and you are unable to reject the null hypothesis.
3. *p-value method:* The p-value $= 0.123 > \alpha = 0.05$ and you are unable to reject the null hypothesis.

You reach the same conclusion no matter which of the three methods you use, as will always be the case.

EXAMPLE I-10
Find the solution to Example I-9 using Excel.

SOLUTION
The Excel solution to the problem is shown in Fig. I-11.

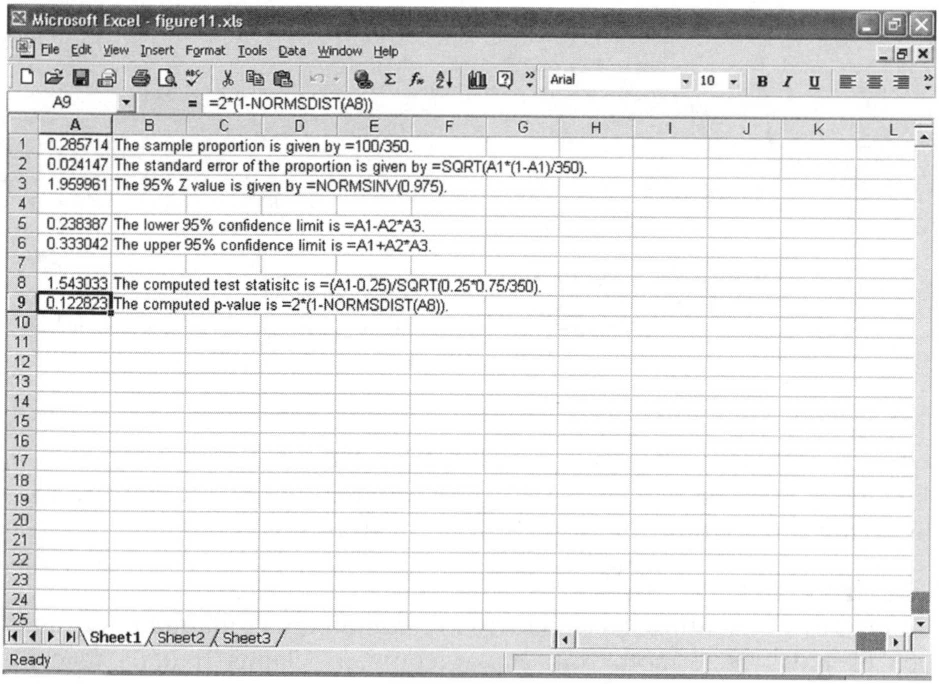

Fig. I-11.

I-4 Inferences About a Population Variance or Standard Deviation

Suppose a sample of size n is taken from a normal distribution and the sample variance, S^2, is computed. (σ^2 is the population variance.) Then $(n-1)S^2/\sigma^2$ has a chi-square distribution with $(n-1)$ degrees of freedom. A chi-square distribution with 9 degrees of freedom is shown in Fig. I-12 to give an idea of the shape of this distribution.

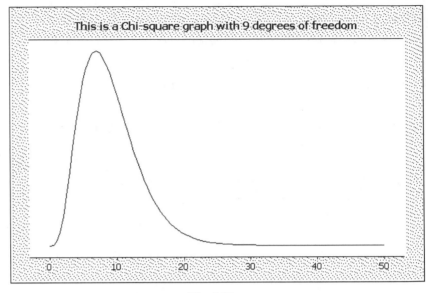

Fig. I-12.

To test H_0: $\sigma^2 = \sigma_0^2$ versus one of the alternatives $\sigma^2 < \sigma_0^2$, $\sigma^2 > \sigma_0^2$, or $\sigma^2 \neq \sigma_0^2$, the test statistic $(n-1)S^2/\sigma_0^2$ is used. This test statistic has a chi-square distribution with $n-1$ degrees of freedom.

EXAMPLE I-11
Companies often utilize machines to fill containers containing substances such as milk, beer, motor oil, etc. A company claims that its machines fill 1-liter containers of motor oil with a standard deviation of less than 2 milliliters. A sample of 10 containers filled by the machine contained the following amounts: 999.01, 1000.78, 1001.02, 998.78, 1000.98, 999.25, 1000.56, 998.75, 1001.78, and 999.76. The null hypothesis is H_0: $\sigma^2 = 4$ versus the research hypothesis H_a: $\sigma^2 < 4$ and $\alpha = 0.05$.

SOLUTION
Figure I-13 shows the test in an Excel worksheet. This test is not directly available in Minitab.

EXAMPLE I-12
The point-spread error is the difference between the game outcome and the point spread that odds-makers establish for games. Table I-3 gives the point-spread error for 100 recent NFL games. Set a 95% confidence interval on the standard deviation.

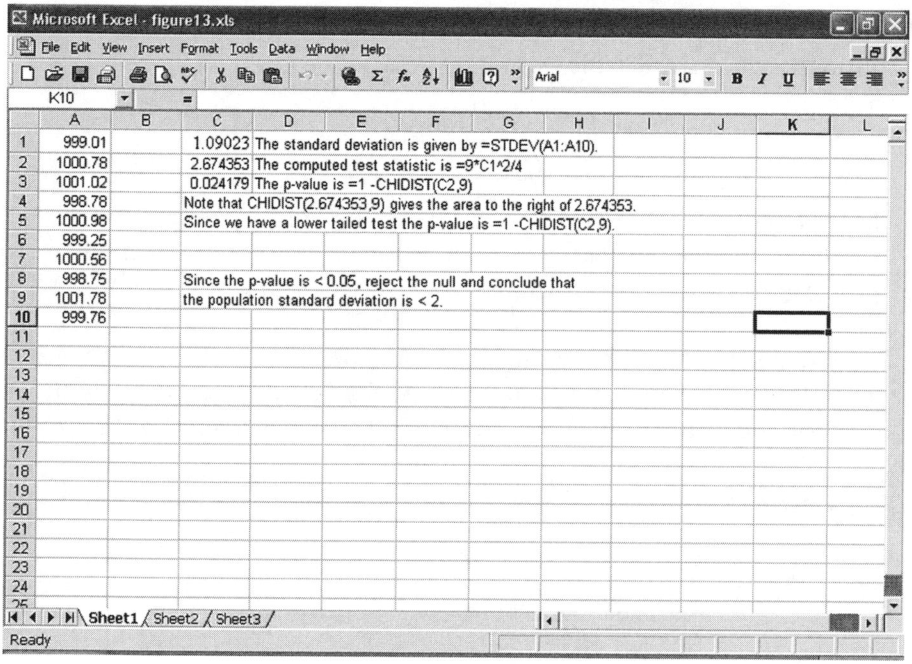

Fig. I-13.

Table I-3 Point spread errors for 100 recent NFL games.

1	3	2	−1	5	−2	0	4	0	4
1	−2	−7	−3	−6	−7	4	2	−3	4
−5	−2	−1	−6	−5	−4	5	−1	0	2
5	−5	1	4	−7	1	−3	0	−2	0
5	−1	−7	−6	−6	−7	−7	−4	−4	0
3	−7	0	−4	−1	−2	2	3	−2	0
0	−7	−7	1	−7	4	−1	4	4	2
−2	−6	0	−6	−2	−7	−7	−7	1	−6
−3	1	−5	5	−3	1	−5	2	1	−6
−4	1	−5	−6	−6	−6	5	−1	5	−2

SOLUTION

A confidence interval for the population variance is

$$\left(\frac{(n-1)S^2}{\chi^2_{\alpha/2}}, \frac{(n-1)S^2}{\chi^2_{1-\alpha/2}}\right)$$

The symbol $\chi^2_{\alpha/2}$ represents a value from the chi-square distribution with $n-1$ degrees of freedom having $\alpha/2$ area to its right; $\chi^2_{1-\alpha/2}$ is a value from the chi-square distribution with $n-1$ degrees of freedom having $1-\alpha/2$ area to its right.

The Excel solution is shown in Fig. I-14. The 95% confidence interval for the population variance is $(11.62566, 20.35125)$ and the 95% confidence interval for the population standard deviation is $(3.409643, 4.511236)$.

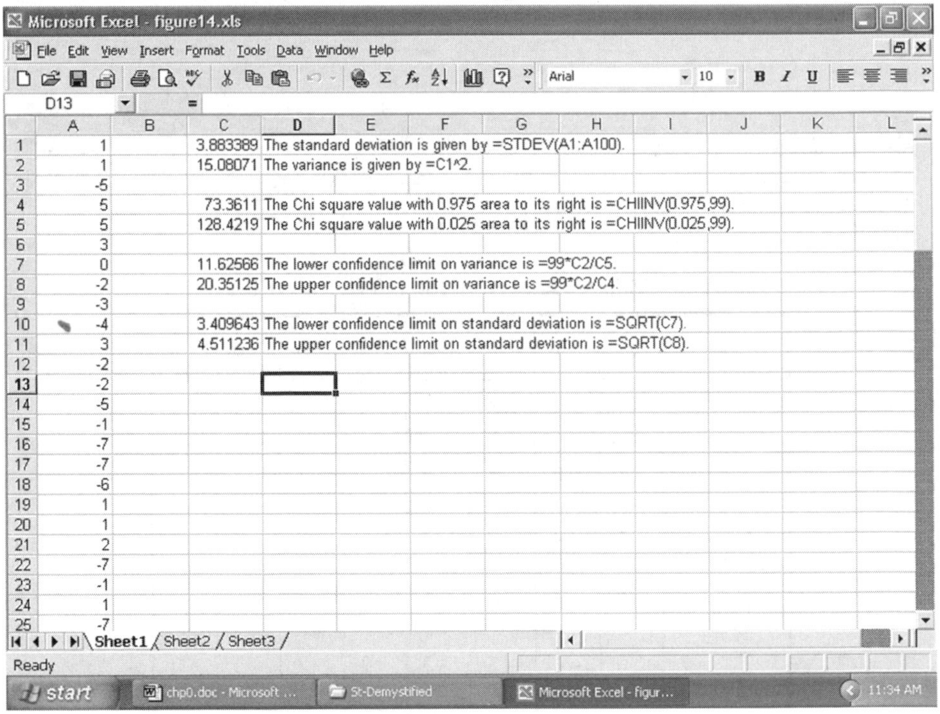

Fig. I-14.

This chapter has given a review of the basics of statistical inference concerning a single population parameter. Confidence intervals and tests of hypotheses have been discussed for means, proportions, and variances. Some of the fundamental ideas involved in statistical thinking have been discussed.

I-5 Using Excel and Minitab to Construct Normal, Student t, Chi-Square, and F Distribution Curves

There are four basic continuous distributions corresponding to test statistics that are used in this book for doing statistical inference. They are the standard normal, the student t, the chi-square, and the F distributions. We need to be able to find areas under these distribution curves. This means that we must be able to construct the probability density curve for the test statistic. In order to determine when some occurrence is unusual we need to be able to relate that occurrence to some value of the test statistic and compute the p-value related to it. Suppose we are doing a large sample test concerning a mean. The value $\bar{x} = 15$ for a sample that we have collected. The null hypothesis states that $\mu = 10$. When we put our sample information and the null hypothesis together, we come up with a test statistic value of $Z = 4.5$. We need to be able to compute the p-value $= 2P(Z \geq 4.5) = 6.80161\text{E} - 06$. This represents an area under the standard normal distribution. We will show how to construct the curves and find areas under the curves in this section.

EXAMPLE I-13
Construct a normal curve, using Excel, describing the heights of adult males with a mean equal to 5 ft 11 inches or 71 inches and a standard deviation equal to 2.5 inches. Construct the curve from 3 standard deviations below the mean to 3 standard deviations above the mean. (Theoretically the curve extends infinitely in both directions.)

SOLUTION
The Excel solution is shown in Fig. I-15. Numbers from 63.5 to 78.5 are entered into column A and =NORMDIST(A1,71,2.5,0) is entered into B1. A click-and-drag is performed on both columns. In the NORMDIST function, the 0 in the fourth position tells Excel to calculate the height of the curve at the number in the first position. Suppose we wished to know the percent of males who are taller than 6 ft 3 inches. The expression =NORMDIST(75,71,2.5,1) in any cell gives 0.9452. This is the percent that are shorter than 75 inches. The 1 in the fourth position of the NORMDIST function tells Excel to accumulate the area from 75 to the left. There would be $1 - 0.9452 = 0.0548$ or 5.48% that are taller than 75 inches.

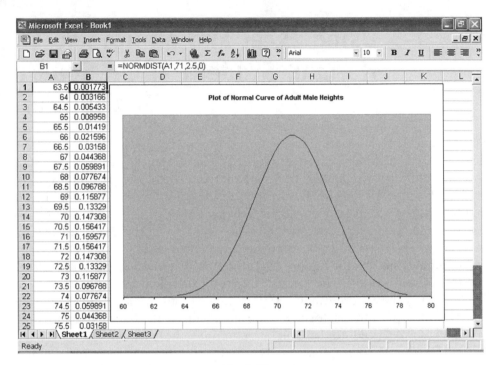

Fig. I-15.

The function =NORMSDIST(Z) returns the cumulative function for the standard normal distribution. The two functions NORMSINV and NORMINV are the inverse functions for NORMSDIST and NORMDIST. For example, suppose we wished to find the male height such that 90% were that tall or shorter. We would request that height as =NORMINV (0.9,71,2.5). The answer is found to be 74.2 inches.

EXAMPLE I-14
Use Minitab to find (a) the value of α for the rejection region $Z > 2.10$ and (b) the two-tailed rejection region that corresponds to $\alpha = 0.15$. Construct a curve to illustrate what you are doing.

SOLUTION
First enter the numbers from -4 to 4 with 0.1 intervals between the numbers. The pull-down **Calc ⇒ Probability Distributions ⇒ Normal** gives a dialog box which is filled as shown in Fig. I-16. This dialog box will calculate the y values for the standard normal curve and place them in column C2. The probability density is checked. This causes the heights to be computed for the curve. The worksheet is shown in Fig. I-17, with the coordinates of

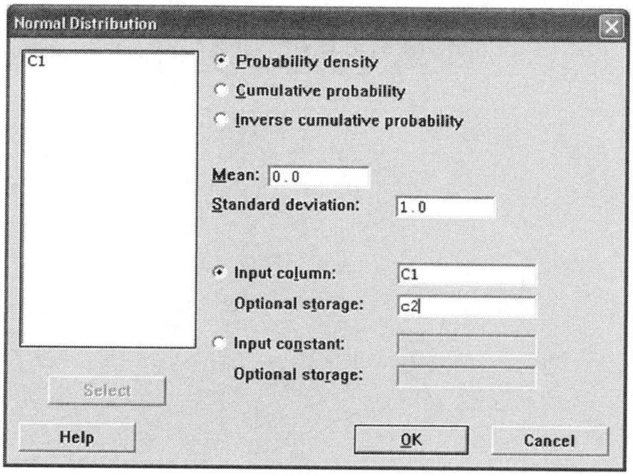

Fig. I-16.

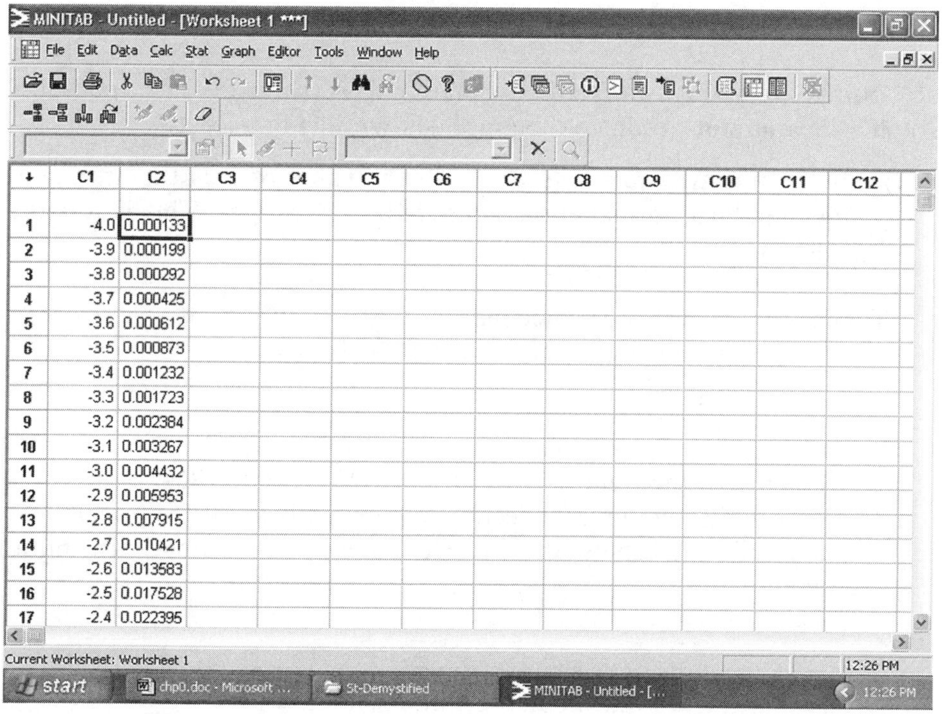

Fig. I-17.

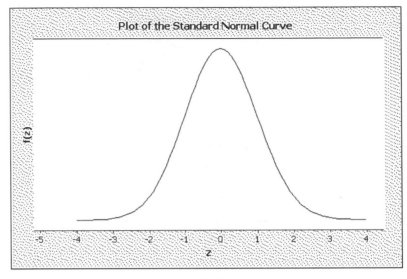

Fig. I-18.

the points on the standard normal curve shown. The pull-down **Graph ⇒ Scatterplot** produces the graph shown in Fig. I-18.

(a) If the rejection region is $Z > 2.10$, the value of α is the area under the standard normal curve from 2.10 to the right. This is found as follows.

Cumulative Distribution Function
```
Normal with mean=0 and standard deviation=1

x          P(X<=x)
2.1        0.982136
```

The area to the right of 2.1 is $1 - 0.982136 = 0.017864$.

(b) To find the two-tailed rejection region with $\alpha = 0.15$, put 0.075 in each tail and apply the inverse cumulative normal distribution function.

Inverse Cumulative Distribution Function
```
Normal with mean=0 and standard deviation=1

P(X<=x)         x
0.075           −1.43953
```

The two-tailed rejection region is $|Z| > 1.43953$.

Plots of the student t, Chi-square, and F distributions are all made in a similar manner using Minitab. First of all construct the (x, y) coordinates on the curves using the pull-down **Calc $\Rightarrow$ Probability Distributions $\Rightarrow$ normal, t, Chi-square, or F**. After the coordinates on the curve are calculated, the pull-down **Graph $\Rightarrow$ Scatterplot** is used to plot the curves. The graphs in Example I-14 were constructed using this technique.

EXAMPLE I-15

Suppose a two-tailed test of a mean using a sample of size 10 gives a test statistic value of $t = 2.10$. Draw a t-curve illustrating the p-value, and find the p-value.

SOLUTION

The shaded region in Fig. I-19 is the p-value.

To find the p-value, we find the area in the left tail and double it. The pull-down **Calc $\Rightarrow$ Probability Distributions $\Rightarrow$ t** gives the dialog box shown in Fig. I-20. We fill it as shown.

Cumulative Distribution Function

```
Student's t distribution with 9 DF

x              P(X<=x)
-2.1           0.0325591
```

The above output is the area to the left of -2.1. We double it and get the p-value to be 0.065.

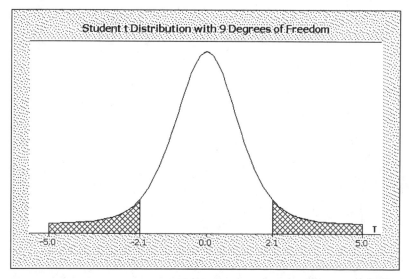

Fig. I-19.

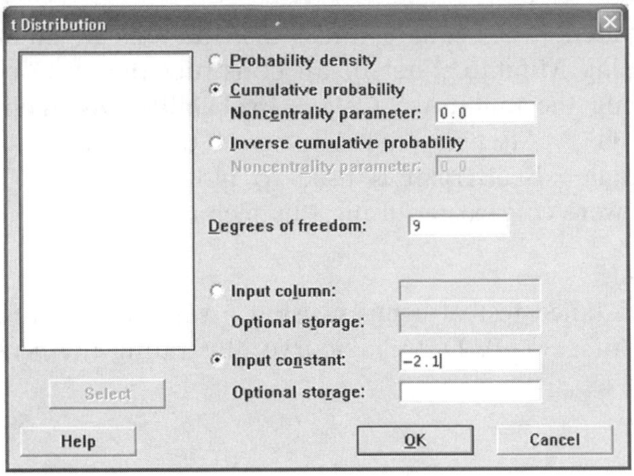

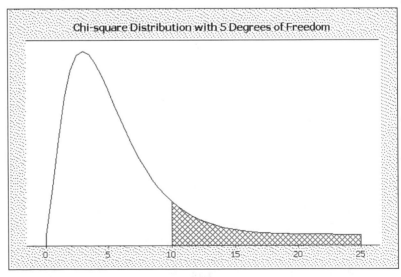

Fig. I-20.

EXAMPLE I-16

An upper-tailed hypothesis test of a population variance is conducted and the computed test statistic is equal to 10. The test statistic is computed from a small sample of size $n = 6$. Draw a chi-square curve illustrating the p-value. Calculate the p-value.

SOLUTION

The graph illustrating the p-value is shown in Fig. I-21, the p-value being the shaded area. First use the chi-square distribution to find the area to the left of 10.

Fig. I-21.

> **Note:** It can be shown that the mean of the chi-square distribution is equal to its degrees of freedom.

Cumulative Distribution Function

```
Chi-Square with 5 DF

x              P(X<=x)
10             0.924765
```

The p-value is $1 - 0.924765 = 0.075$.

I-6 Exercises for Introduction

1. Give the values for α for the following rejection regions: (a) $Z < -1.88$; (b) $Z > 2.45$; (c) $|Z| > 2.12$.
2. Give the rejection regions in terms of Z for the following α values and research hypotheses: (a) $\alpha = 0.075$ and H_a: $\mu > \mu_0$; (b) $\alpha = 0.12$ and H_a: $\mu < \mu_0$; (c) $\alpha = 0.17$ and H_a: $\mu \neq \mu_0$.
3. Find the p-values that are determined by the following computed test statistics and research hypotheses: (a) $Z = -1.87$ and H_a: $\mu < \mu_0$; (b) $Z = 2.45$ and H_a: $\mu > \mu_0$; (c) $Z = -2.56$ and H_a: $\mu \neq \mu_0$.
4. For the following p-values and α values, give your conclusion: (a) p-value $= 0.045$ and $\alpha = 0.05$; (b) p-value $= 0.35$ and $\alpha = 0.10$; (c) p-value $= 0.001$ and $\alpha = 0.01$.
5. Fifty adult onset diabetics took part in a study. One of the questions asked was the number of planned hours of exercise per week that the diabetic participated in. The data is shown in Table I-4.

   After consulting the following Minitab output, answer the questions.

```
Test of mu = 5 vs not = 5
The assumed standard deviation = 3.4

Variable  N  Mean     StDev    SE Mean  95% CI               Z      P
hours     50 4.92000  3.39772  0.48083  (3.97759, 5.86241)   -0.17  0.868
```

   (a) What is the null and the research hypothesis?
   (b) What is your conclusion at $\alpha = 0.05$ and why, using the confidence interval method?
   (c) What is your conclusion at $\alpha = 0.05$ and why, using the classical method?

Table I-4 Hours of planned exercise per week.

0	7	5	5	0
3	10	3	0	5
5	3	0	0	10
10	0	10	3	3
7	10	10	3	3
3	3	10	5	10
3	3	3	10	0
5	9	5	0	3
9	5	5	3	3
10	5	9	5	5

(d) What is your conclusion at $\alpha = 0.05$ and why, using the p-value method?

6. A sample of size 10 is used to test a hypothesis about a mean. Give the values for α for the following rejection regions: (a) $t > 2.262$; (b) $t < -3.250$; (c) $|t| > 4.297$.

7. Give the rejection region in terms of t for the following α values, research hypothesis, and sample size $= 15$: (a) $\alpha = 0.05$ and H_a: $\mu > \mu_0$; (b) $\alpha = 0.10$ and H_a: $\mu < \mu_0$; (c) $\alpha = 0.01$ and H_a: $\mu \neq \mu_0$.

8. Find the p-values that are determined by the following computed test statistic and research hypothesis for a sample of 5: (a) $t = -2.132$ and H_a: $\mu < \mu_0$; (b) $t = 2.776$ and H_a: $\mu > \mu_0$; (c) $t = -3.747$ and H_a: $\mu \neq \mu_0$.

9. The temperatures of twenty patients who had contacted a rare type of flu are shown in Table I-5.

The Minitab t-test for a single mean is as follows:

```
Test of mu = 105 vs not = 105

Variable     N    Mean     StDev   SE Mean   95% CI                    T     P
temperature  20   103.650  2.996   0.670     (102.248, 105.052)  -2.02  0.058
```

Table I-5 Temperatures of twenty flu patients.

107	107	101	104	102
106	107	100	106	100
106	100	107	106	100
101	105	107	101	100

    (a)   What is the null and the research hypothesis?

    (b)   What is your conclusion at $\alpha = 0.05$ and why, using the confidence interval method?

    (c)   What is your conclusion at $\alpha = 0.05$ and why, using the classical method?

    (d)   What is your conclusion at $\alpha = 0.05$ and why, using the p-value method?

10.   In a recent *USA Today* snapshot, it was reported that 75% of office workers make up to 100 copies per week, 13% make from 101 to 1000, 5% make more than 1000, and 7% make none. In a similar survey of 150 workers, it was reported that 8% made more than 1000 copies. A Minitab analysis of the survey results is as follows:

```
Test of p = 0.05 vs p not = 0.05

Sample X   N    Sample p   95% CI                 Z-Value  P-Value
1       12  150  0.080000   (0.036585, 0.123415) 1.69      0.092
```

    (a)   What is the null and the research hypothesis?

    (b)   What is your conclusion at $\alpha = 0.05$ and why, using the confidence interval method?

    (c)   What is your conclusion at $\alpha = 0.05$ and why, using the classical method?

    (d)   What is your conclusion at $\alpha = 0.05$ and why, using the p-value method?

    (e)   Is the sample large enough for the normal approximation to be valid?

11.   A sample of monthly returns on a portfolio of stocks, bonds, and other investments is given in Table I-6. Find a 95% confidence interval on σ.

Table I-6

2500	1870	2250	1990	2350
2400	2130	1980	2340	2550

After looking over the following Excel output, give your answer.

	A	B	C	D	E	F	G	H	I	J
1	2500		233.4381	The standard deviation is given by =STDEV(A1:A10)						
2	2400		54493.33	The variance is given by =C1^2						
3	1870									
4	2130		2.700389	The Chi-square value with 0.975 area to its right is =CHIINV(0.975,9)						
5	2250		19.02278	The Chi-square value with 0.025 area to its right is =CHIINV(0.025,9)						
6	1980									
7	1990		25781.72	The lower confidence limit on variance is =9*C2/C5						
8	2340		181618.3	The upper confidence limit on variance is =9*C2/C4						
9	2350									
10	2550		160.5669	The lower confidence limit on standard deviation is =SQRT(C7)						
11			426.167	The upper confidence limit on standard deviation is =SQRT(C8)						
12										

12. A large sample test procedure was performed to test H_0: $\mu = 750$ versus H_a: $\mu < 750$ at $\alpha = 0.01$, where μ represents the mean number of children for which a school nurse is responsible. The study involved 300 nurses and the study results were: $\bar{x} = 715$, standard error of the mean = 20. Compute the test statistic, the p-value, and give your conclusion for the hypothesis test. Use Excel to do your computations.

13. Repeat problem 12 for a sample size of 20 with everything else remaining the same. Give your answers, using Excel, to do your computations.

14. A survey was taken of 350 nurses and one question that was asked was "Do you think the 12-hour shift that you work affects your job performance?" There were 237 "yes" responses. Set a 95% confidence interval on the population proportion that would respond yes.

15. An industrial process dyes materials. It is important that the color be uniform. A color-o-meter records colors at various times in the process. The sample variance of 20 of the readings is 3.45. Set a 95% confidence interval on the population standard deviation.

16–19. Find the following areas under the given curves (Figures I-22–I-25 respectively). **(Remember: The curve extends infinitely in both directions.)**

16.

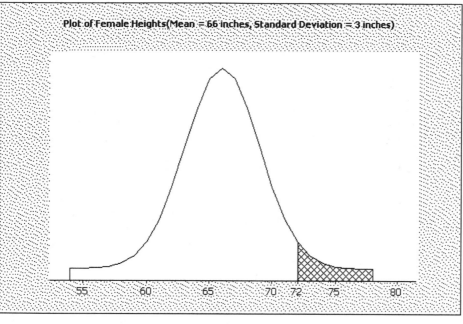

Fig. I-22.

17.

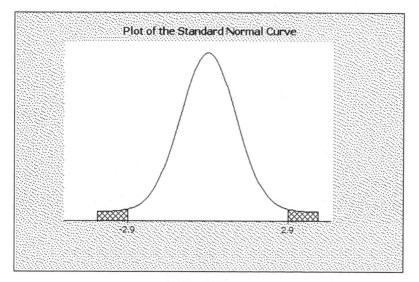

Fig. I-23.

18.

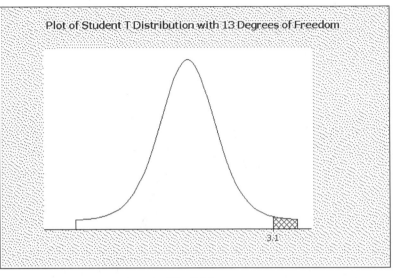

Plot of Student T Distribution with 13 Degrees of Freedom

3.1

Fig. I-24.

19.

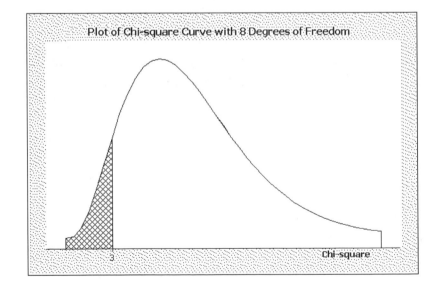

Plot of Chi-square Curve with 8 Degrees of Freedom

3

Chi-square

Fig. I-25.

I-7 Introduction Summary

LARGE SAMPLE INFERENCES ABOUT A SINGLE MEAN

"Large sample," when making inferences about a single population mean, is taken to mean that the sample size $n > 30$. The sample mean, $\bar{x}$, is a *point estimate* or a single numerical estimate for the population mean, μ. The *interval estimate* is a better estimate because it gives the reliability associated with the estimate. A $(1 - \alpha)$ confidence interval estimate for μ is $(\bar{x} \pm z_{\alpha/2}s/\sqrt{n})$.

A hypothesis test about μ is of the following form. The null hypothesis H_0: $\mu = \mu_0$ versus one of the three research hypotheses: H_a: $\mu < \mu_0$, $\mu > \mu_0$, or $\mu \neq \mu_0$. The test statistic is $Z = (\bar{x} - \mu_0)/(s/\sqrt{n})$. The statistic, Z, has a standard normal distribution. The statistical test is performed by one of three methods: the *classical method*, the *p-value method*, or the *confidence interval method*.

The Minitab pull-down **Stat** $\Rightarrow$ **Basic Statistics** $\Rightarrow$ **1-sample Z** is used to set a confidence interval or test a hypothesis about μ when your sample is large ($n > 30$).

SMALL SAMPLE INFERENCES ABOUT A SINGLE MEAN

The test about a single mean for a small sample is based on a test statistic that has a student t distribution with degrees of freedom equal to $n - 1$, where n is the sample size. It is assumed that the sample is taken from a normally distributed population. This assumption needs to be checked. If the population has a skew or is bi- or trimodal, for example, a non-parametric test should be used rather than the t-test. To estimate the population mean for small samples, use the following confidence interval.

$$\bar{x} \pm t_{\alpha/2}\frac{s}{\sqrt{n}}$$

To test the null hypothesis that the population mean equals some value μ_0, calculate the following test statistic:

$$T = \frac{\bar{x} - \mu_0}{s/\sqrt{n}}$$

Note that this test statistic is computed just as it was for a large sample. The difference is that, for small samples, it does not have a standard normal distribution (Z). It tends to be more variable because of the small sample. The T distribution is very similar to the Z distribution. It is bell-shaped and centers at zero. If a T curve with $n-1$ degrees of freedom and a Z curve are plotted together, it can be seen that the T curve has a standard deviation larger than one.

The Minitab pull-down **Stat** $\Rightarrow$ **Basic Statistics** $\Rightarrow$ **1-sample t** is used to set a confidence interval or test a hypothesis about μ when your sample is small.

LARGE SAMPLE INFERENCES ABOUT A SINGLE POPULATION PERCENT OR PROPORTION

The test statistic $Z = (\hat{p} - p_0)/\sqrt{p_0 q_0/n}$ is used to test the null hypothesis H_0: $p = p_0$ against any one of the alternatives $p < p_0$, $p > p_0$, or $p \neq p_0$. The hypothesized percent in the population with the characteristic of interest is p_0 and the percent in the sample of size n having the characteristic is $\hat{p}$. The normal approximation to the binomial distribution is used and is valid if $np_0 > 5$ and $nq_0 > 5$. If estimation is performed, the confidence interval is of the form $\hat{p} \pm z_{\alpha/2}\sqrt{\hat{p}\hat{q}/n}$. For the confidence interval to be valid, the sample size must be large enough so that $n\hat{p} > 5$ and $n\hat{q} > 5$. Almost all surveys meet this sample size requirement so that the standard normal approximation described above will be valid.

The Minitab pull-down **Stat** $\Rightarrow$ **Basic Statistics** $\Rightarrow$ **1-proportion** is used to set a confidence interval or test a hypothesis about p when your sample is large.

INFERENCES ABOUT A SINGLE POPULATION STANDARD DEVIATION OR VARIANCE

To test H_0: $\sigma^2 = \sigma_0^2$ versus one of the alternatives $\sigma^2 < \sigma_0^2$, $\sigma^2 > \sigma_0^2$, or $\sigma^2 \neq \sigma_0^2$, the test statistic $(n-1)S^2/\sigma_0^2$ is used. This test statistic has a chi-square distribution with $n-1$ degrees of freedom. A confidence interval for the population variance is

$$\left(\frac{(n-1)S^2}{\chi_{\alpha/2}^2}, \ \frac{(n-1)S^2}{\chi_{1-\alpha/2}^2} \right)$$

The symbol $\chi_{\alpha/2}^2$ represents a value from the chi-square distribution with $n-1$ degrees of freedom having $\alpha/2$ area to its right, and $\chi_{1-\alpha/2}^2$ is a value from the chi-square distribution with $n-1$ degrees of freedom having $1-\alpha/2$ area to its right.

CHAPTER 1

Inferences Based on Two Samples

1-1 Inferential Statistics

Inferential statistics, also called *statistical inference*, is the process of generalizing from *statistics* calculated on samples to *parameters* calculated on populations. In this chapter, we will be concerned with using two sample means to make inferences about two population means, using two sample proportions to make inferences about two population proportions, and using two sample standard deviations to make inferences about two population standard deviations. In particular, we will calculate the following statistics on samples taken from two populations: $\bar{X}_1$, $\bar{X}_2$, S_1, S_2, $\hat{P}_1$, and $\hat{P}_2$. These are the symbols used to represent the mean of the sample from the first population, the mean of the sample from the second population, the standard deviation of the sample from the first population, the standard deviation of the sample from the second population, the proportion in the sample from the first population having a particular characteristic, and the proportion in the sample from the second population having a particular characteristic, respectively. The corresponding measures made on the populations are called parameters. They are represented by the symbols μ_1, μ_2, σ_1, σ_2, P_1, and P_2, respectively.

Estimation and **testing hypothesis** are the two types of statistical inference that occur in real-world problems. For example, we may use $\bar{X}_1 - \bar{X}_2$ to estimate $\mu_1 - \mu_2$ or test a hypothesis about $\mu_1 - \mu_2$, or we may use $\hat{P}_1 - \hat{P}_2$ to estimate $P_1 - P_2$ or test a hypothesis about $P_1 - P_2$, or we may use S_1/S_2 to make inferences about σ_1/σ_2. The methods for doing this will be illustrated in the next four sections.

1-2 Comparing Two Population Means: Independent Samples

Purpose of the test: The purpose of the test is to compare the means of two populations when independent samples have been chosen.

Assumptions: The two independent samples are selected from normal populations having equal variances.

EXAMPLE 1-1
What is the relationship between the mean heights of males and females? We suspect that, on the average, males are taller than females. Our research hypothesis is stated as follows: H_a: $\mu_{\text{male}} > \mu_{\text{female}}$. The null hypothesis is H_0: $\mu_{\text{male}} = \mu_{\text{female}}$. The test is conducted at $\alpha = 0.05$.

SOLUTION
Independent samples of $n_1 = 10$ male heights and $n_2 = 10$ female heights are selected. The data are entered for the males in column C1 and for the females in column C2 of the Minitab worksheet (Fig. 1-1). The variable names are entered at the top of the columns. The pull-down menu **Stat** $\Rightarrow$ **Basic Statistics** $\Rightarrow$ **2-sample t** gives Fig. 1-2, the dialog box, which is filled out as shown. The software calculates the value of

$$t = \frac{\bar{X}_1 - \bar{X}_2 - 0 \text{ (the null hypothesis value for } \mu_1 - \mu_2)}{S_{\text{pooled}}\sqrt{\dfrac{1}{n_1} + \dfrac{1}{n_2}}}$$

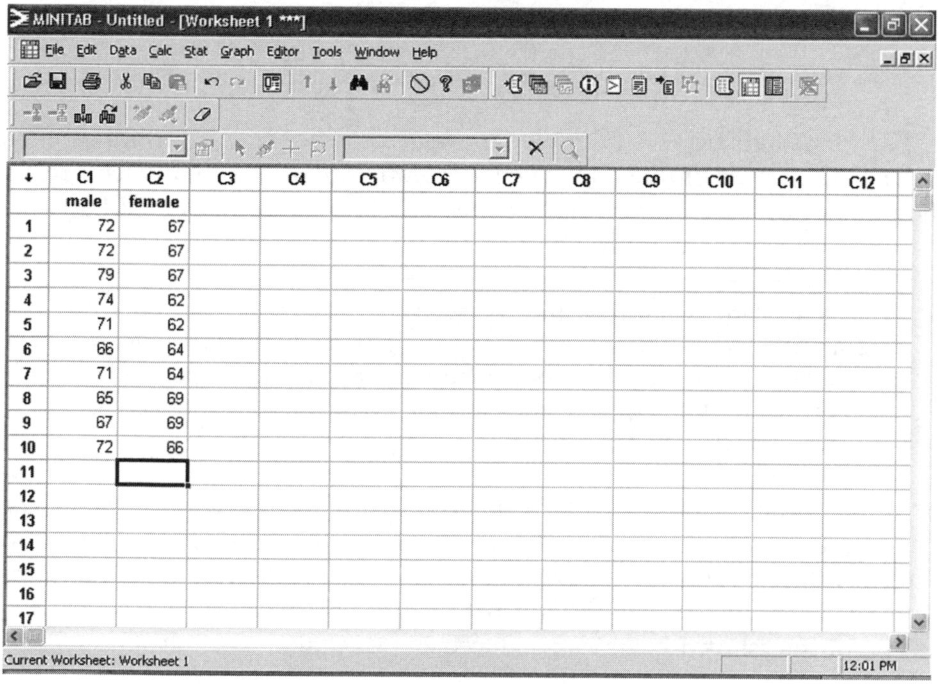
MINITAB - Untitled - [Worksheet 1 ***]
File Edit Data Calc Stat Graph Editor Tools Window Help
C1 C2 C3 C4 C5 C6 C7 C8 C9 C10 C11 C12
male female
1 72 67
2 72 67
3 79 67
4 74 62
5 71 62
6 66 64
7 71 64
8 65 69
9 67 69
10 72 66
11
12
13
14
15
16
17
Current Worksheet: Worksheet 1
12:01 PM

Fig. 1-1.

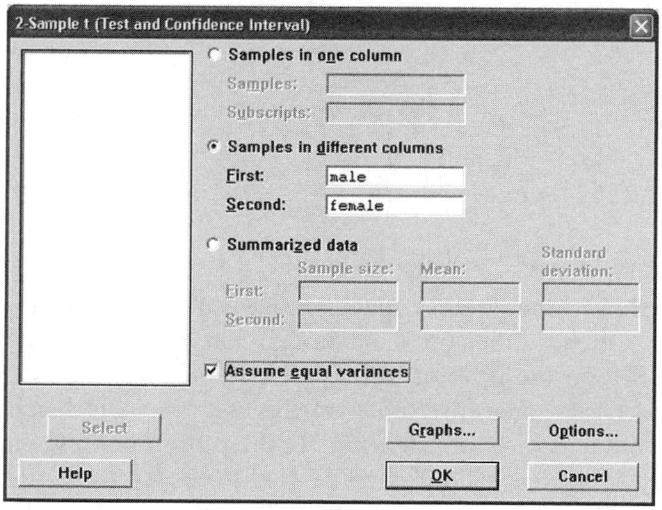
2-Sample t (Test and Confidence Interval)
Samples in one column
Samples:
Subscripts:
Samples in different columns
First: male
Second: female
Summarized data
Sample size:
Mean:
Standard deviation:
First:
Second:
Assume equal variances
Select
Graphs...
Options...
Help
OK
Cancel

Fig. 1-2.

where

$$S^2_{\text{pooled}} = \frac{(n_1 - 1)S_1^2 + (n_2 - 1)S_2^2}{n_1 + n_2 - 2}$$

The test statistic, t, is known to have a student t distribution with $n_1 + n_2 - 2$ degrees of freedom. The two sample variances are pooled because the population variances are assumed equal. The subscript 1 identifies males and the subscript 2 identifies females.

> **Note:** Zero is put into the test statistic for $\mu_1 - \mu_2$ because we always assume the null hypothesis to be true. This is the same as saying that $\mu_1 = \mu_2$. If we obtain an unusually large or small value for the test statistic, then we reject the null hypothesis and accept the research hypothesis.

In the output below, the value for this test statistic is shown to be $t = 3.46$. The computer program calculates the area to the right of 3.46 under the student t distribution curve having $(10 + 10 - 2) = 18$ degrees of freedom and finds that area to be 0.001. This is the p-value for the test.

The output shown below is produced by Minitab.

Two-sample T for male vs female

	N	Mean	StDev	SE Mean
male	10	70.94	4.09	1.3
female	10	65.67	2.52	0.80

```
Difference = mu male − mu female
Estimate for difference: 5.27
95% lower bound for difference: 2.63
T-Test of difference = 0 (vs >): T-Value = 3.46 P-Value = 0.001
DF = 18
Both use Pooled StDev = 3.40
```

The output shows that the estimated difference is 5.27 inches; that is, the males are 5.27 inches taller on the average. The p-value is 0.001. Since this p-value is less than the level of significance (0.05), we reject the null hypothesis and infer that males are taller on the average. Another interpretation of the p-value is in order at this point. If the null hypothesis is true, that is that males and females are the same height on the average, there is a probability of 0.001, or 1 chance out of 1000, that sample means, based on samples of size 10, could be this far apart or further.

EXAMPLE 1-2
Solve Example 1-1 using Excel.

SOLUTION

The Excel solution is shown in Figs. 1-3, 1-4, and 1-5. The pull down **Tools ⇒ Data Analysis** produces the data analysis dialog box shown in Fig. 1-3. The **t-Test: Two-Sample Assuming Equal Variances** test is chosen. The corresponding dialog box is filled in as shown in Fig. 1-4. The output shown in Fig. 1-5 is created from the data in columns A and B. The number on line 11 is the *p*-value, equal to 0.001667. This small *p*-value indicates that the null hypothesis of equal means should be rejected and the conclusion reached that on the average men are taller than women.

Note: If large samples are available ($n_1 > 30$ and $n_2 > 30$), the normality assumption and the equal variances assumption may be dropped, and the test statistic, Z, be used to test the hypothesis, where

$$Z = \frac{\bar{x}_1 - \bar{x}_2 - (\text{the value of } \mu_1 - \mu_2 \text{ stated in the null})}{\sqrt{\dfrac{s_1^2}{n_1} + \dfrac{s_2^2}{n_2}}}$$

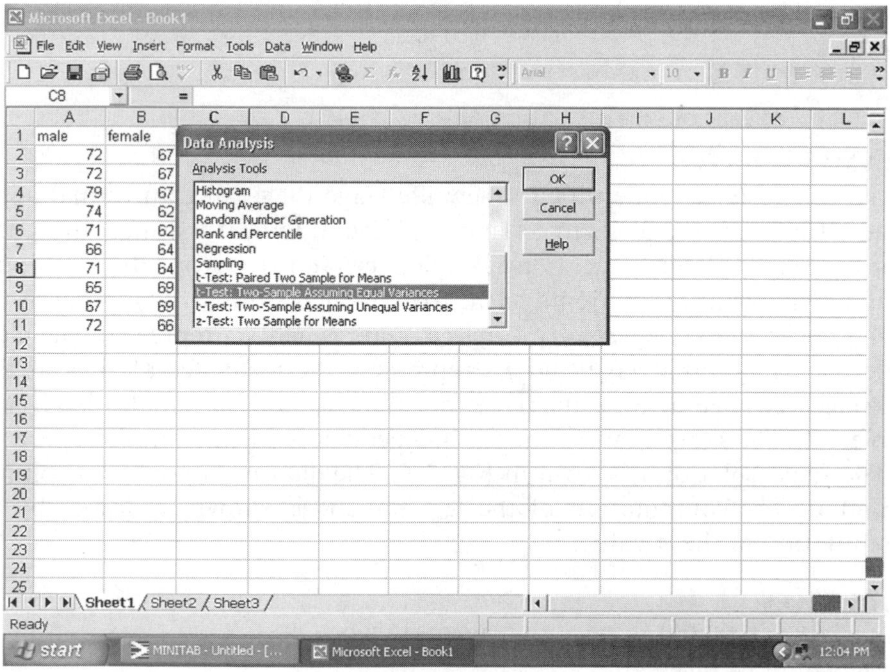

Fig. 1-3.

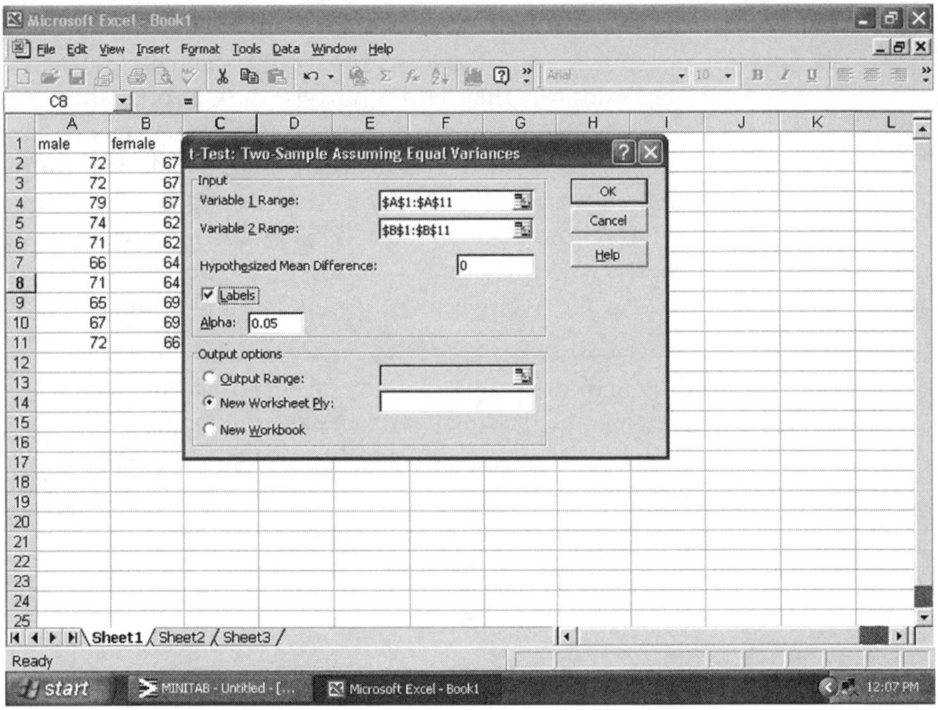

Fig. 1-4.

EXAMPLE 1-3

Social scientists have identified a new life stage they call transitional adulthood. It lasts from age 18 to 34 and has several indicators: median age for first marriage is later, education takes longer, and the proportion of young adults living with their parents has increased. In a study, the hypothesis H_0: $\mu_1 - \mu_2 = 2$ years versus H_a: $\mu_1 - \mu_2 > 2$ years, where μ_1 represents the mean male age at first marriage and μ_2 represents the mean female age at first marriage. The age at first marriage for 50 males and 50 females is given in Table 1-1. The samples were chosen independently of one another.

The Excel solution is shown in Fig. 1-6. The data is entered into columns A and B. The computation of the test statistic is shown, followed by the computation of the p-value.

SOLUTION

The p-value $1.14799E - 09 = 0.0000000014799$ indicates that the null hypothesis would be rejected.

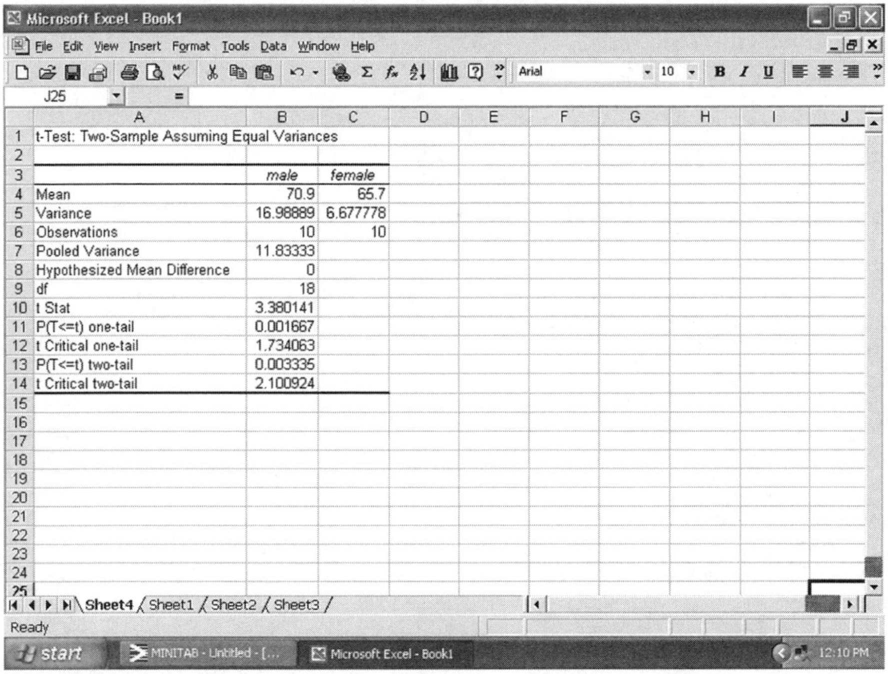

Fig. 1-5.

Table 1-1   Age at first marriage for 50 males and 50 females.

Males					Females				
38	34	30	25	36	23	30	27	34	23
32	25	30	26	31	25	17	23	27	24
21	30	28	30	31	23	22	26	29	25
32	29	31	27	29	28	26	27	25	23
33	32	31	31	35	23	26	23	26	21
31	30	36	31	30	22	24	24	23	23
32	30	29	31	25	31	26	23	21	25
30	32	32	29	32	26	25	22	22	25
28	33	31	31	33	24	19	28	27	23
27	30	23	33	26	24	23	22	20	27

	A	B	C	D	E	F	G	H	I	J	K
1	male	female									
2	38	23		30.24	AVERAGE(A2:A51)						
3	32	25		24.5	AVERAGE(B2:B51)						
4	21	23		3.242322171	STDEV(A2:A51)						
5	32	28		3.011881235	STDEV(B2:B51)						
6	33	23		5.975923083	(D2-D3-2)/SQRT(D4^2/50+D5^2/50)						
7	31	22									
8	32	31		1.14799E-09	1-NORMSDIST(D6)						
9	30	26									
10	28	24									
11	27	24									
12	34	30									
13	25	17									
14	30	22									
15	29	26									
16	32	26									
17	30	24									
18	30	26									
19	32	25									
20	33	19									
21	30	23									
22	30	27									
23	30	23									
24	28	26									
25	31	27									

Fig. 1-6.

1-3 Comparing Two Population Means: Paired Samples

Purpose of the test: The purpose of the test is to compare the means of two samples when dependent samples have been chosen.

Assumptions: The two samples are dependent and the differences in the sample values are normally distributed.

EXAMPLE 1-4

Suppose we are interested in determining whether a diet is effective in producing weight loss in overweight individuals. In fact, suppose we believe the diet will result in more than a 10 pounds weight loss over a six-month period. The 16 overweight individuals are weighed at the beginning of the experiment and again at the six-month period. The research hypothesis is H_a: $\mu_{diff} > 10$. The null hypothesis is H_0: $\mu_{diff} \leq 10$ or H_0: $\mu_{diff} = 10$. (Note: If the data causes us to reject $\mu_{diff} = 10$ and accept H_a: $\mu_{diff} > 10$, it would also require us to reject $\mu_{diff} \leq 10$ and accept H_a: $\mu_{diff} > 10$.) We decide before starting the experiment to run the test at $\alpha = 0.01$.

SOLUTION

The data are entered into the worksheet as shown in Fig. 1-7. The one-sample t test is performed on Diff. The pull down **Stat** $\Rightarrow$ **Basic Statistics** $\Rightarrow$ **1-Sample t** is used to analyze the Diff values. If $\bar{d}$ is the mean sample difference and S_d is the standard deviation of the sample differences, then

$$t = \frac{\bar{d} - 0 \text{ (assuming the mean population difference is 0)}}{S_d/\sqrt{n}}$$

has a student t distribution with $n - 1 = 16 - 1 = 15$ degrees of freedom. Minitab computes the value of this test statistic and finds its value to be $t = 0.18$. The area to the right of 0.18 on the student t curve with 15 degrees of freedom is found to be 0.43. This is the p-value for the test. This does not lead us to reject the null hypothesis.

One-Sample t: Diff

Test of mu = 10 vs mu > 10

Variable	N	Mean	StDev	SE Mean
Diff	16	10.52	11.56	2.89

Variable	95.0% Lower Bound	T	P
Diff	5.45	0.18	0.430

The output shows an upper tail test, a sample size $n = 16$, a mean weight loss of 10.52 pounds, a standard deviation of 11.56, a standard error equal to 2.89 pounds, a computed test statistic of $T = 0.18$, and a p-value $= 0.43$. This large p-value shows no evidence for rejecting the null. The evidence does not cause us to believe that the diet results in more than a 10 pound weight loss.

EXAMPLE 1-5

Give the Excel solution for Example 1-4.

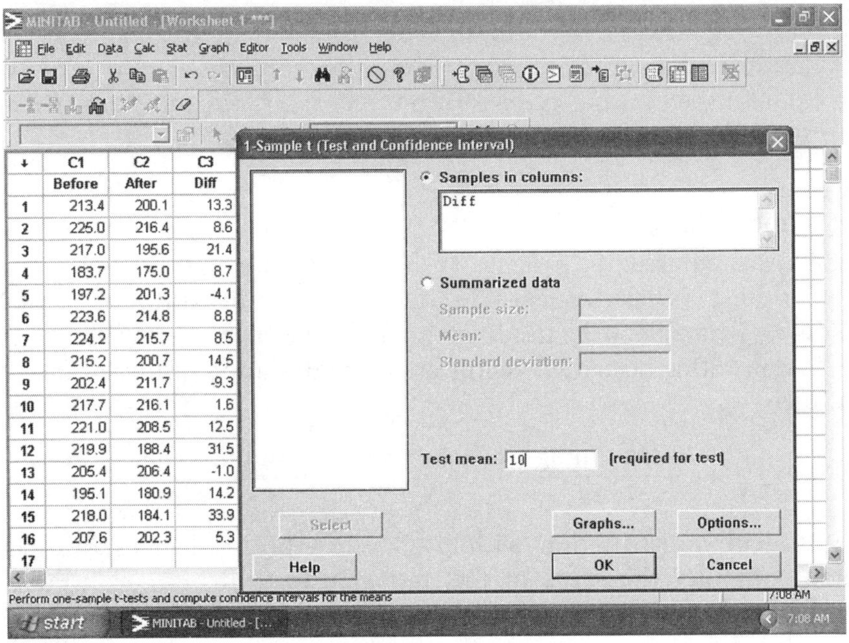

Fig. 1-7.

The MINITAB worksheet contains the following data:

	C1	C2	C3
	Before	After	Diff
1	213.4	200.1	13.3
2	225.0	216.4	8.6
3	217.0	195.6	21.4
4	183.7	175.0	8.7
5	197.2	201.3	-4.1
6	223.6	214.8	8.8
7	224.2	215.7	8.5
8	215.2	200.7	14.5
9	202.4	211.7	-9.3
10	217.7	216.1	1.6
11	221.0	208.5	12.5
12	219.9	188.4	31.5
13	205.4	206.4	-1.0
14	195.1	180.9	14.2
15	218.0	184.1	33.9
16	207.6	202.3	5.3

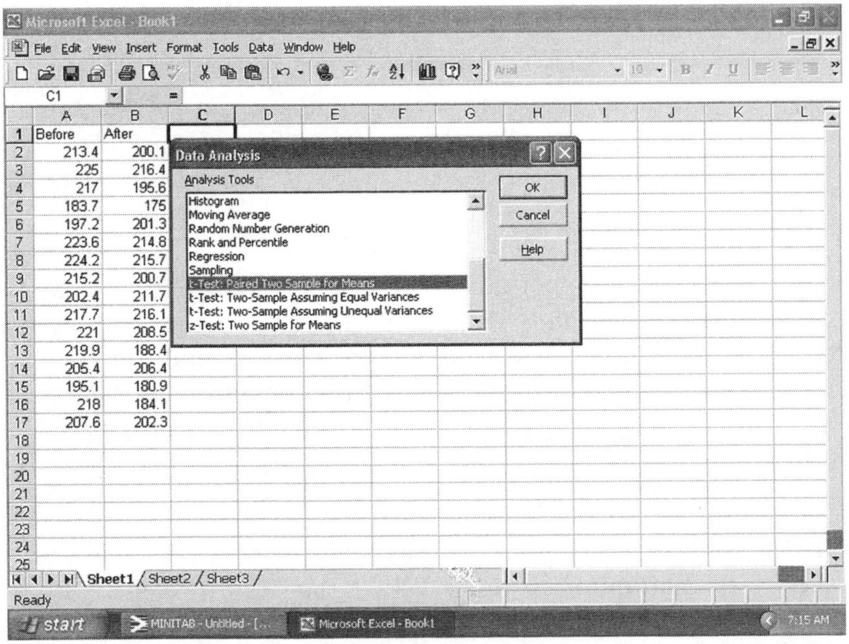

The Excel worksheet contains the following data:

	A	B
1	Before	After
2	213.4	200.1
3	225	216.4
4	217	195.6
5	183.7	175
6	197.2	201.3
7	223.6	214.8
8	224.2	215.7
9	215.2	200.7
10	202.4	211.7
11	217.7	216.1
12	221	208.5
13	219.9	188.4
14	205.4	206.4
15	195.1	180.9
16	218	184.1
17	207.6	202.3

Fig. 1-8.

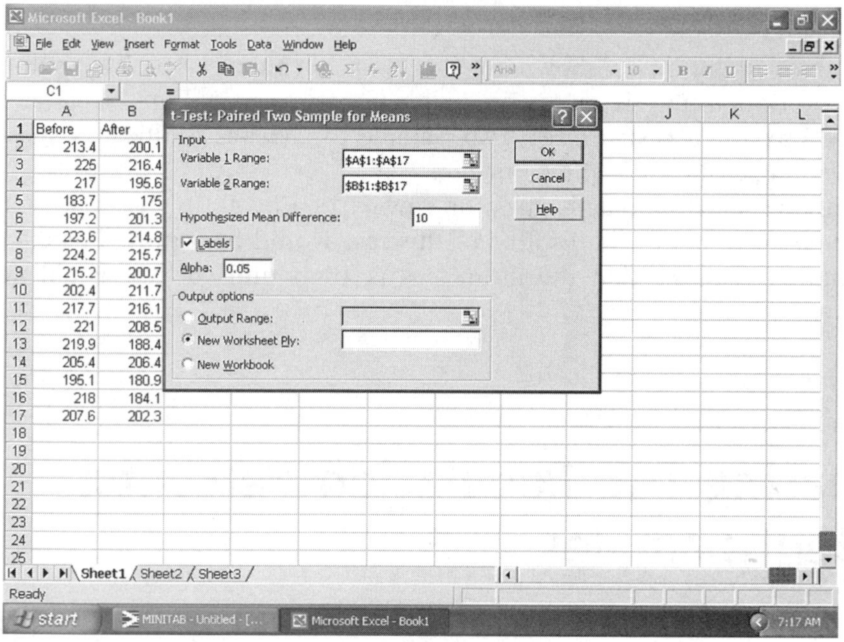

Fig. 1-9.

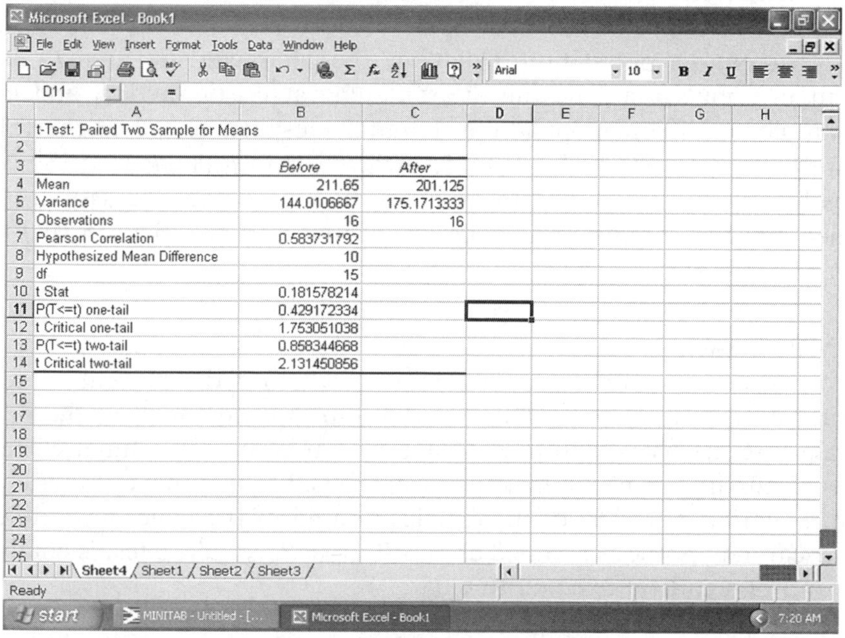

Fig. 1-10.

SOLUTION

The Excel solution is shown in Figs. 1-8 through 1-10. The pull-down **Tools** ⇒ **Data Analysis** gives the dialog box shown in Fig. 1-8. The Excel dialog box for **t-Test: Paired Two Sample for Means** is filled as shown in Fig. 1-9. The output is shown in Fig. 1-10.

The one-tailed *p*-value is given on row 11 of Fig. 1-10. Again, it is shown to be equal to 0.43. The *p*-value of this test would be reported as 0.43 and the null hypothesis that the mean loss is 10 pounds or less could not be rejected.

1-4 Comparing Two Population Percents: Independent Samples

Purpose of the test: The purpose of the test is to compare two populations with respect to the percent in the populations having a particular characteristic.

Assumptions: The samples are large enough so that the normal approximation to the binomial distribution holds in both populations. That is, if P_1 is the percent in population 1 having the characteristic and P_2 is the percent in population 2 having the same characteristic, and n_1 and n_2 are the sample sizes from populations 1 and 2, respectively, then the following are true: $n_1 P_1 \geq 5$, $n_1(1 - P_1) \geq 5$, $n_2 P_2 \geq 5$, and $n_2(1 - P_2) \geq 5$. Since the values for P_1 and P_2 are unknown, check to see whether the assumption is satisfied for the sample results; i.e., if $n_1 \hat{p}_1 \geq 5$, $n_1(1 - \hat{p}_1) \geq 5$, $n_2 \hat{p}_2 \geq 5$, and $n_2(1 - \hat{p}_2) \geq 5$.

EXAMPLE 1-6

Suppose we are interested in determining whether the percent of female Internet users who have visited a chat room is different from the percent of male Internet users who have visited a chat room. Two hundred female and two hundred male Internet users are asked if they have visited a chat room. It is found that 67 of the females and 45 of the males have done so. Our research hypothesis is stated as H_a: $P_{male} \neq P_{female}$. The test is conducted at a level of significance equal to 0.05.

SOLUTION

The test statistic for this test is

$$Z = \frac{\hat{p}_1 - \hat{p}_2 - 0 \text{ (null hypothesis is no difference in proportions)}}{\sqrt{\hat{p}\hat{q}\left(\frac{1}{n_1} + \frac{1}{n_2}\right)}}$$

where $\hat{p} = (x_1 + x_2)/(n_1 + n_2)$ and $\hat{q} = 1 - \hat{p}$. In this case $x_1 = 67$, $x_2 = 45$, $n_1 = n_2 = 200$. The test statistic z has a standard normal distribution.

The following pull-down sequence is given. **Stat $\Rightarrow$ Basic Statistics $\Rightarrow$ 2 Proportions.** The dialog box is filled in as shown in Fig. 1-11. The output is as follows.

Test and CI for Two Proportions

Sample	X	N	Sample p
1	67	200	0.335000
2	45	200	0.225000

Estimate for p(1)−p(2): 0.11
95% CI for p(1)−p(2): (0.0226606, 0.197339)
Test for p(1)−p(2)=0 (vs not=0): Z=2.47   P-Value=0.014

The output tells us the following. Thirty three point five percent of the female sample visited a chat room and twenty two point five percent of the male sample did so. The estimated difference in the population proportions is 11%. A 95% confidence interval for the difference is a low of 2.27% and a high of 19.73%. The value for test statistic z is found by the software to be 2.47. Because a two-tailed hypothesis is being tested, the area under the standard normal curve to the right of 2.47 is found and its value is doubled. This is the p-value. The p-value is 0.014 and the null hypothesis is rejected at the $\alpha = 0.05$ level. We conclude that a greater percentage of females visit a chat room.

1-5 Comparing Two Population Variances

Purpose of the test: The purpose of the test is to determine whether there is equal variability in two populations.

Assumptions: It is assumed that independent samples are selected from two populations that are normally distributed.

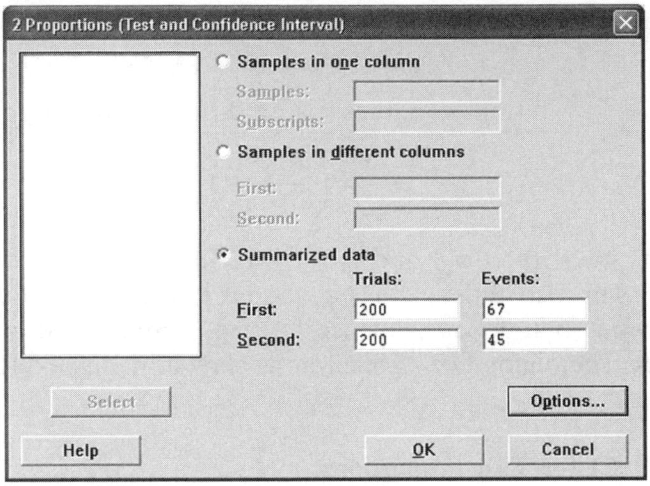

Fig. 1-11.

EXAMPLE 1-7
In order to compare the variability of two kinds of structural steel, an experiment was undertaken in which measurements of the tensile strength of each of twelve pieces of each type of steel were taken. The units of measurement are 1000 pounds per square inch. The research hypothesis $H_a: \sigma_1^2 \neq \sigma_2^2$ is to be tested at $\alpha = 0.01$.

SOLUTION
The data for the samples are entered into the Minitab worksheet as shown in Fig. 1-12. The pull-down menu **Stat ⇒ Basic Statistics ⇒ 2 Variances** gives Fig. 1-13.

The output is as follows.

Test for Equal Variances
```
Level1        steel 1
Level2        steel 2
ConfLvl       95.0000
```

Bonferroni confidence intervals for standard deviations

Lower	Sigma	Upper	N	Factor Levels
3.04101	4.49709	8.31292	12	steel 1
1.30994	1.93715	3.58084	12	steel 2

F-Test (normal distribution)

Test Statistic: 5.389
P-Value: 0.009

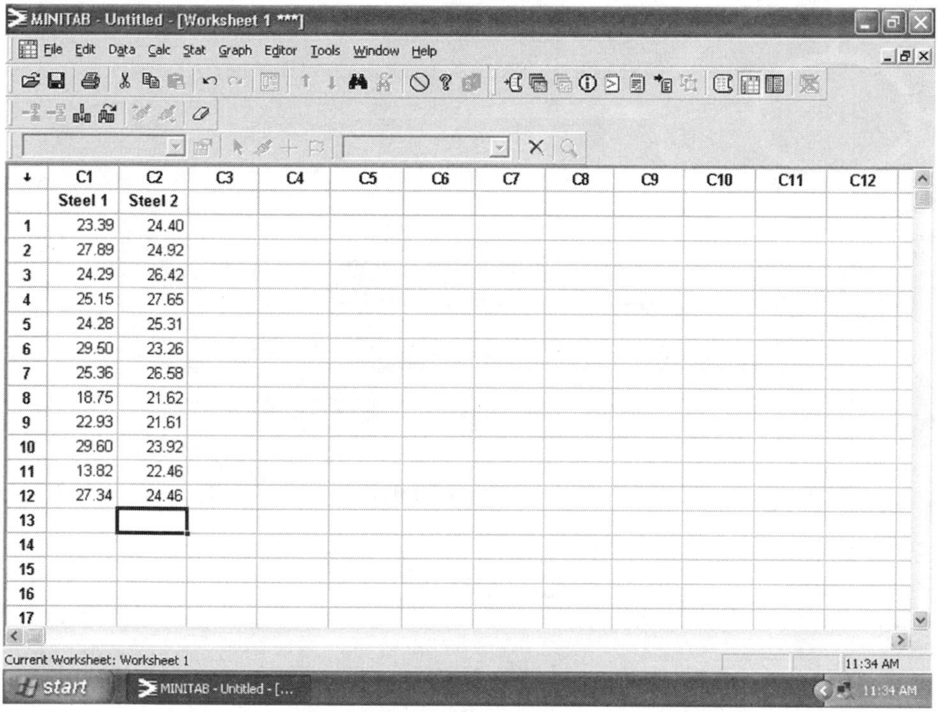

Fig. 1-12.

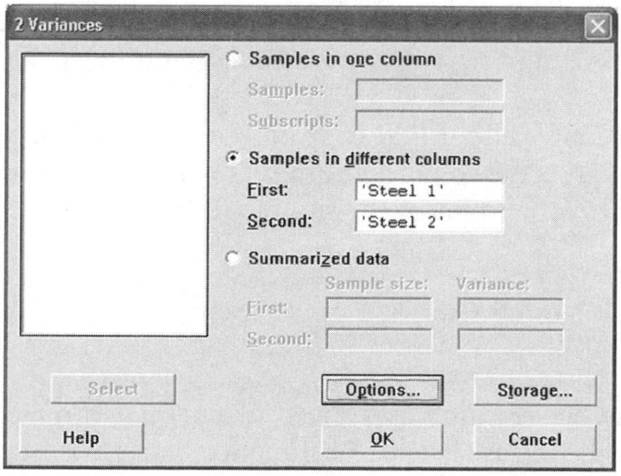

Fig. 1-13.

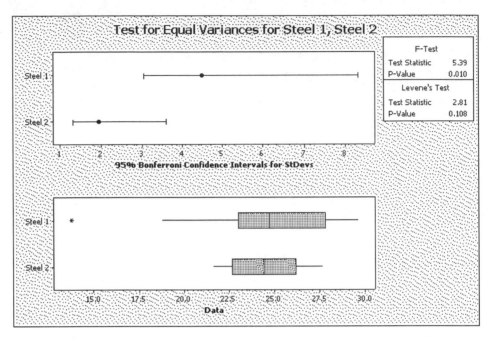

Fig. 1-14.

The output gives a 95% confidence interval for the population 1 standard deviation as $(3.04, 8.31)$ and for the population 2 standard deviation as $(1.31, 3.58)$. The test statistic is $F = (S_1^2/S_2^2) = 5.389$ and the p-value, based on the F-distribution, is 0.009. On the basis of this p-value we would reject the null hypothesis and accept the research hypothesis that the variances are unequal.

There is additional output that is given as shown in Fig. 1-14. The upper part of Fig. 1-14 gives the confidence interval in graphic form. The lower part of the figure gives a box plot of the sample data.

EXAMPLE 1-8
Solve Example 1-7 using Excel.

SOLUTION
The Excel test of the research hypothesis of unequal variances is as follows. The **Paste** function is used, and the function category **Statistical** and the function name **FTEST** are chosen (see Fig. 1-15). Clicking OK produces the dialog box shown in Fig. 1-16, which is filled in as shown.

The output that is given is the p-value, which is seen to be 0.0095. Compare this to the value given in the Minitab output in Fig. 1-14.

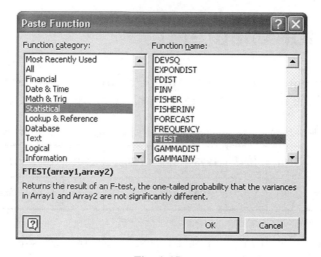

Fig. 1-15.

Fig. 1-16.

1-6 Exercises for Chapter 1

1. In a recent *USA Today* snapshot entitled "Auto insurance bill to jump," the following averages (Table 1-2) were reported.

Table 1-2   Average annual consumer spending on auto insurance.

1995	1996	1997	1998	1999	2000	2001	2002	2003
$668	$691	$707	$704	$683	$687	$723	$784	$885

A research study compared 1995 and 2004 by sampling 50 auto insurance payments for these two years. The results are shown in Table 1-3.

Test H_0: $\mu_1 - \mu_2 = \$200$ versus H_a: $\mu_1 - \mu_2 > \$200$ at $\alpha = 0.05$ by giving the following.

Table 1-3   Comparison of auto insurance bills for 1995 and 2004.

1995					2004				
714	685	658	663	691	896	908	881	873	855
659	655	668	686	625	908	896	870	923	917
686	704	647	669	651	933	901	862	883	939
662	668	644	681	730	907	904	918	925	878
712	642	687	658	670	897	902	921	880	895
647	645	720	658	674	865	859	860	907	870
675	668	688	693	670	859	872	863	915	935
625	685	702	683	627	914	943	872	869	877
685	656	656	684	668	936	908	907	867	894
695	624	667	683	681	864	935	868	941	882

    (a)   Give the difference between the two sample means.
    (b)   Give the standard error of the difference in the sample means.
    (c)   Give the test statistic.
    (d)   Give the p-value and your conclusion.

2.   A research study was conducted to compare two methods of teaching statistics in high school. One method, called the traditional method, presented the course without the use of computer software. The other method, referred to as the experimental method, taught the course and utilized Excel software extensively. The scores made on a common comprehensive final by the students in both sections are shown in Table 1-4. Test the hypothesis that the experimental method produced higher scores on the average. Use $\alpha = 0.05$. Do the assumptions appear satisfied?

Table 1-4 Comparison of two methods of teaching statistics using independent samples.

Traditional	82	85	82	73	72	82	73	79	71	86	90	98	86	77	81
Experimental	78	83	96	89	82	83	68	84	83	76	83	89	90	85	77

3.   In order to compare fertilizer A and fertilizer B, a paired experiment was conducted. Ten two-acre plots were chosen throughout the region and the two fertilizers were randomly assigned to the plots. That is, it was randomly decided which of the fertilizers was applied to the northern one-acre plot and the other fertilizer was applied to the southern one-acre plot. The yields of wheat per acre were recorded and are given in Table 1-5. Test the research hypothesis that there is a difference in average yield, depending on the fertilizer, at $\alpha = 0.01$. Answer the following questions in performing your analysis. Form your differences by subtracting fertilizer B yield from fertilizer A yield.

Table 1-5 Comparison of wheat yields due to different fertilizers on paired plots.

Plot	1	2	3	4	5	6	7	8	9	10
A	60	65	79	55	75	60	69	108	77	88
B	57	60	70	60	70	65	59	101	67	86

   (a)   What is the value of the average difference?
   (b)   What is the standard deviation of the differences?
   (c)   What is the computed value of the test statistic?
   (d)   What is the p-value for the test, and what is your conclusion?

4.   In a study designed to determine whether taking aspirin reduces the chance of having a heart attack, 11,000 male physicians took aspirin on a regular basis and 11,000 male physicians took a placebo on a regular basis. The researchers determined whether the physician suffered a heart attack over a five-year period. Test $H_0: (p_1 - p_2) = 0$ versus $H_a: (p_1 - p_2) < 0$ (where $p_1 =$ proportion of men who regularly take aspirin who suffer a heart attack and $p_2 =$ proportion of men who do not take aspirin regularly who suffer a heart attack). Table 1-6 gives the number having a heart attack in both groups.

Table 1-6   Comparison of heart attack rates for two groups.

	Sample size	$x =$ number
Aspirin	$n_1 = 11,000$	$x_1 = 105$
Placebo	$n_2 = 11,000$	$x_2 = 189$

Analyze the experiment by answering the following questions.

   (a)   Give the values for $\hat{p}_1$ and $\hat{p}_2$.
   (b)   Give the point estimate for $P_1 - P_2$.
   (c)   Give the value of the test statistic.
   (d)   Give the p-value and your conclusion for the study.

5.   An experimenter wants to compare the metabolic rates of mice subjected to different drugs. The weights of the mice affect their metabolic rates so the researcher wants to obtain mice that are fairly homogeneous with respect to weight. She obtains samples from two different companies that sell mice for research and obtains their weights in ounces. The results are shown in Table 1-7.
   Test that the weights of mice sold by the two companies have different variances at $\alpha = 0.05$. Analyze the experiment by answering the following questions.

   (a)   Give the standard deviations of the two samples.
   (b)   Give the value of the F statistic.
   (c)   Give the p-value and your conclusion for the test.

Table 1-7   Weights of mice sold by two different companies.

Company A					Company B				
4.12	4.13	4.47	3.62	4.27	4.10	4.44	3.71	4.57	4.21
3.96	4.22	4.28	4.55	4.70	3.60	4.00	4.45	3.80	3.82
4.29	4.25	4.31	4.12	4.04	4.00	4.16	4.11	4.34	3.97
4.76	4.08	4.26	4.39	4.47	3.36	4.19	3.63	4.01	4.12

6. Two types of chicken feed were compared. One hundred chicks were fed Diet 1 and 100 were fed Diet 2. The summary statistics for a 3-month period were as follows: Diet 1: mean gain $= 1.45$, variance $= 0.75$; Diet 2: mean gain $= 1.84$, variance $= 0.61$. The null hypothesis is that the population means are equal and the research hypothesis is that the means are not equal. Give the test statistic and the p-value that accompanies it. State your conclusion at $\alpha = 0.01$.

7. Suppose the sample sizes in problem 6 were five each. Everything else remains the same. Give the test statistic and the p-value that accompanies it. State your conclusion at $\alpha = 0.01$.

8. An experiment was conducted to determine the effects of weight loss on blood pressure. The blood pressure of 25 patients was determined at the beginning of an experiment. After the patients had lost 10 pounds, their blood pressure was checked again and the difference was formed as: difference $=$ before minus after. None of the patients were on blood pressure medicine. The 25 differences had a mean of 7.3 and a standard deviation of 2.5. Was the drop in blood pressure significant at $\alpha = 0.05$? To answer the question, calculate the test statistic and give the p-value.

9. A survey of teenagers was taken and it was determined how many spent 20 or more hours in front of a TV. The results from a survey of 500 males and 500 females were as follows: males: 70 of 500; females: 40 of 500. Find the p-value for the research hypothesis that the percents differ.

10. An instructor wishes to compare the variances of final exam scores that were given in a large-enrollment algebra course for two consecutive years. Random samples of 25 of the test scores were selected from the two years with the following

results: year 1: variance $= 25.3$; year 2: variance $= 33.5$. Give the test statistic for testing $H_{\mathrm{a}}: \sigma_1^2 \neq \sigma_2^2$ and the p-value that goes with the test statistic.

1-7 Chapter 1 Summary

COMPARING TWO POPULATION MEANS: INDEPENDENT SAMPLES

Test Statistic

$$t = \frac{\bar{X}_1 - \bar{X}_2 - 0 \text{ (the null value for } \mu_1 - \mu_2)}{S_{\text{pooled}}\sqrt{\dfrac{1}{n_1} + \dfrac{1}{n_2}}}$$

where

$$S_{\text{pooled}}^2 = \frac{(n_1 - 1)S_1^2 + (n_2 - 1)S_2^2}{n_1 + n_2 - 2}$$

Minitab Pull-down

Stat $\Rightarrow$ **Basic Statistics** $\Rightarrow$ **2-sample t**

Excel Pull-down

Tools $\Rightarrow$ **Data Analysis** followed by t-Test: Two-Sample Assuming Equal Variances

---

Note: If large samples are available ($n_1 > 30$ and $n_2 > 30$), the normality assumption and the equal variances assumption may be dropped and the test statistic, Z, be used to test the hypothesis, where

$$Z = \frac{\bar{x}_1 - \bar{x}_2 - \text{(the value of } \mu_1 - \mu_2 \text{ stated in the null)}}{\sqrt{\dfrac{s_1^2}{n_1} + \dfrac{s_2^2}{n_2}}}$$

COMPARING TWO POPULATION MEANS: PAIRED SAMPLES

Test Statistic

$$t = \frac{\bar{d} - 0 \text{ (assuming the mean population difference is 0)}}{S_d/\sqrt{n}}$$

Minitab Pull-down

Stat $\Rightarrow$ **Basic Statistics** $\Rightarrow$ **1-Sample t** is used to analyze the Diff values.

Excel Pull-down

Tools $\Rightarrow$ **Data Analysis** followed by t-Test: Paired Two Sample for Means.

COMPARING TWO POPULATION PERCENTS: INDEPENDENT SAMPLES

Test Statistic

$$Z = \frac{\hat{p}_1 - \hat{p}_2 - 0 \text{ (null hypothesis is no difference in proportions)}}{\sqrt{\hat{p}\hat{q}\left(\frac{1}{n_1} + \frac{1}{n_2}\right)}}$$

where $\hat{p} = (x_1 + x_2)/(n_1 + n_2)$ and $\hat{q} = 1 - \hat{p}$.

Minitab Pull-down

Stat $\Rightarrow$ **Basic Statistics** $\Rightarrow$ **2 Proportions**

Excel Pull-down

There is no Excel pull-down. You would need to compute the test statistic and then use Excel to compute the *p*-value.

COMPARING TWO POPULATION VARIANCES

Test Statistic

$$F = \frac{S_1^2}{S_2^2}$$

Minitab Pull-down

Stat $\Rightarrow$ Basic Statistics $\Rightarrow$ 2 Variances

Excel Solution

The paste function is used and the function category **Statistical** is chosen; then the function name **FTEST** is chosen.

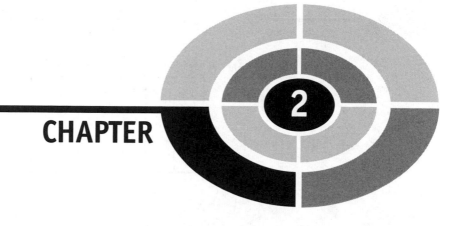

Analysis of Variance: Comparing More Than Two Means

2-1 Designed Experiments

We are interested in the relationship between the amount of fertilizer applied and the yield of wheat. We are interested in three levels of fertilizer: low, medium, and high. There are eighteen similar plots, numbered 1 through 18, available and the plots are located on an experimental farm at Midwestern University. The fertilizer is called a *factor* and there are three *levels* of interest for this factor. The plots are numbered as shown in Table 2-1.

The fertilizer levels are applied randomly as follows. Six random numbers are chosen between 1 and 18. The numbers are 1, 13, 8, 10, 16, and 12 and a

Table 2-1 Numbered plots at Midwestern University Farm.

1	2	3	4	5	6
7	8	9	10	11	12
13	14	15	16	17	18

Table 2-2 Random assignment of treatments to experimental units.

L	M	H	M	H	M
H	L	M	L	M	L
L	M	H	L	H	H

low amount of fertilizer is applied to these plots. From the remaining twelve numbers, six more are randomly chosen. They are 2, 14, 9, 4, 11, and 6. A medium amount of fertilizer is applied to plots with these numbers. A high amount is applied to the remaining plots.

L represents a plot with low application of fertilizer, M a plot with medium application, and H a plot with high application. L, M, and H are levels of fertilizer. They are also called *treatments*. Table 2-2 shows the random assignment of treatments to *experimental units* or plots.

The wheat yield is measured for each plot. The wheat yield is called the *response variable*. Suppose the mean yield for plots with a low application of fertilizer is 22.5 bushels, with a medium application of fertilizer it is 30.5 bushels, and with a high application of fertilizer it is 17.0 bushels. What do these sample means indicate about μ_L, μ_M, and μ_H, the population means? We'll answer this question in the next section.

The experimental design used here is called a *one-way design* or a *completely randomized design*. It is also called a *single factor design* with k *levels*. In the present example $k = 3$. The randomization tends to protect us against extraneous factors that may have been overlooked.

Suppose a fertility gradient runs from left to right because the experimental farm is composed of rolling hills. That is, the plots on the right end are different with respect to fertility than the plots on the left end (Table 2-3).

A different design would then be used to account for the fertility blocks, an additional source of variability. This design is referred to as a **randomized**

Table 2-3   Farm with a fertility gradient.

1	2	3	4	5	6
7	8	9	10	11	12
13	14	15	16	17	18

———— Fertility gradient ————→

Table 2-4   Randomized complete block design.

Block 1	Block 2	Block 3	Block 4	Block 5	Block 6
L	H	H	M	L	H
H	M	L	H	M	L
M	L	M	L	H	M

complete block design. In this design, the treatments are randomly assigned within the blocks. Within a block, the three treatments will be exposed to the same fertility level. The three treatment means will still be made up of six observations each. The block means will be made up of three observations each (see Table 2-4).

Suppose the six block means are equal to 15.6, 16.4, 20.7, 22.6, 30.5, and 36.4 bushels. It was probably worthwhile to block since these means are quite different. The randomized complete block design allows a test to be performed on block means as well as treatment means.

The technique used to analyze the means from a designed experiment is called an *analysis of variance table* or *ANOVA table*. Variances of different sources are analyzed to make inferences about the means of the sources. Different designs have different analysis of variance tables.

Consider next a *multi-factor experiment*. Suppose we are not only interested in the effect of the three levels of fertilizer (L, M, and H) on the wheat yield, but we are also interested in the effect of two levels of moisture, low and high (L and H) on the wheat yield. These three levels of fertilizer, called factor A, and two levels of moisture, referred to as factor B, give six factor-level combinations or treatments, as shown in Table 2-5.

Table 2-5   Six factor-level combinations.

Treatment	Fertilizer	Moisture
trt 1	L	L
trt 2	L	H
trt 3	M	L
trt 4	M	H
trt 5	H	L
trt 6	H	H

Table 2-6   3 by 2 factorial completely randomized design.

trt 1	trt 5	trt 3	trt 5	trt 2	trt 4
trt 6	trt 2	trt 6	trt 1	trt 6	trt 5
trt 3	trt 4	trt 1	trt 4	trt 3	trt 2

Suppose the same experimental farm is available. Each of the six treatments can be randomly applied to three plots or experimental units. The experiment is called a *3 by 2 factorial* that has been *replicated* three times. The six treatments have been applied in a completely randomized fashion (see Table 2-6). This design allows a test for the *main effect of fertilizer*, a test for the *main effect of moisture*, and a test for the *interaction of moisture and fertilizer*.

A factorial design in three blocks is shown in Table 2-7. In this case blocks can be tested as well as main effects and interaction.

We have introduced only a few of the many experimental designs that are possible and that are studied in a design course. A statistics master's degree student will often take a full year's course in the design and analysis of experiments. The remaining sections of this chapter will discuss a completely randomized design, a block design, and a factorial design in detail. The sections will show how the ANOVAs are built for each of these designs and how the F distribution can be used to test differences in means.

Table 2-7   3 by 2 factorial in three blocks.

Block 1		Block 2		Block 3	
trt 2	trt 4	trt 5	trt 2	trt 1	trt 5
trt 5	trt 1	trt 1	trt 6	trt 6	trt 2
trt 6	trt3	trt 4	trt 3	trt 3	trt 4

2-2 The Completely Randomized Design

Purpose of the test: The purpose of the test is to compare the means of several populations when independent samples have been chosen. This test can be thought of as an extension of the two independent samples t-test of Section 1-2.

Assumptions: (1) Samples are selected randomly and independently from the p populations. (2) All p population probability distributions are normal. (3) The p population variances are equal.

EXAMPLE 2-1
We wish to compare three brands of golf balls with respect to the distance they travel when hit by a mechanical driver. Five balls of brand A, five of brand B, and five of brand C are driven by the mechanical device in a random order. The distance that each travels is measured and the data are shown in Table 2-8. Can the experimental results allow us to conclude that the mean distances traveled are different for the three brands?

Table 2-8   Distances traveled by three brands of golf balls.

Brand A	Brand B	Brand C
246	243	265
231	246	260
236	243	265
217	235	253
246	235	291

SOLUTION

The data are entered into the Minitab worksheet as shown in Fig. 2-1.

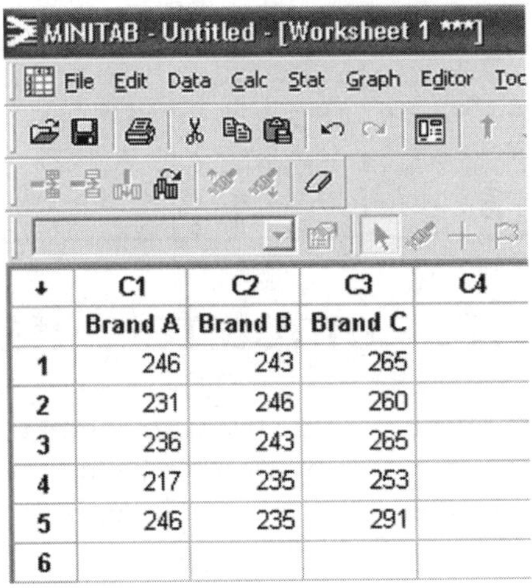

↓	C1	C2	C3	C4
	Brand A	Brand B	Brand C	
1	246	243	265	
2	231	246	260	
3	236	243	265	
4	217	235	253	
5	246	235	291	
6				

Fig. 2-1.

The pull-down **Stat ⇒ ANOVA ⇒ Oneway (unstacked)** gives the dialog box shown in Fig. 2-2. Brand A, Brand B, and Brand C are clicked into the response box. The following output is produced.

One-way ANOVA: Brand A, Brand B, Brand C

```
Analysis of Variance

Source     DF        SS        MS        F        P
Factor      2      2871      1435    11.37    0.002
Error      12      1515       126
Total      14      4386
                                  Individual 95% CIs For Mean
                                  Based on Pooled StDev
Level     N   Mean     StDev    -+- - - -+- - - -+- - - -+- - -
BrandA    5   235.20   12.07    (- - - * - - -)
BrandB    5   240.40    5.08        (- - - * - - -)
BrandC    5   266.80   14.39                      (- - - * - - -)
                                -+- - - -+- - - - -+- - - -+- - -
Pooled StDev=11.24              225      240       255     270
```

The ANOVA output shows that the total variation, as measured by the total sums of squares, is 4386. "Total sums of squares" is nothing but the numerator of the expression for the variance of the fifteen distances.

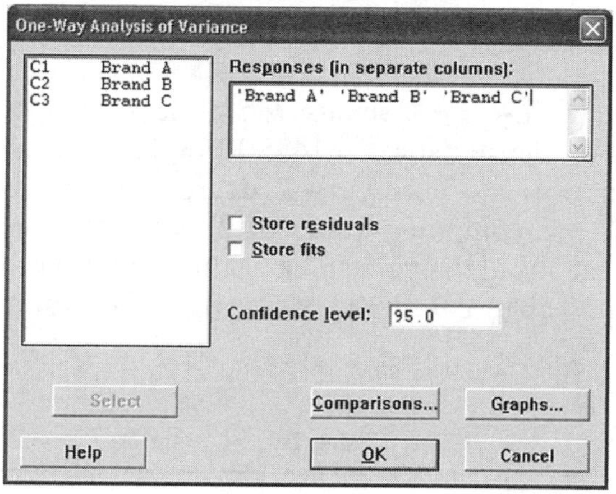

Fig. 2-2.

The total sums of squares $= \sum (x - \bar{\bar{x}})^2 = 4386$, where $\bar{\bar{x}}$ is the mean over all 15 measurements and the total degrees of freedom $= n - 1 = 14$. The total sum of squares is the total variation in the whole data set.

This variation is broken into two parts. The factor sum of squares, also known as the between treatments sums of squares, measures the proximity of the sample means to each other. The factor or between sums of squares $= \sum_{j=1}^{j=3} n_j(\bar{x}_j - \bar{\bar{x}})^2 = 2871$, where $\bar{x}_j$ is the mean for the jth treatment and the treatment degrees of freedom equals one less than the number of treatments $= 3 - 1 = 2$.

The error or within treatments sums of squares is a pooling of the variation within the treatments. It is a pooled measure of the variation within the samples. The error sums of squares or within sums of squares $= (n_1 - 1) S_1^2 + (n_2 - 1)S_2^2 + (n_3 - 1)S_3^2 = 1515$, where S_1^2, S_2^2, and S_3^2 are the sample variances within the three brands. The error degrees of freedom is the difference between the total and factor degrees of freedom $= 14 - 2 = 12$. The following relationship can be shown algebraically:

$$\sum (x - \bar{\bar{x}})^2 = \sum_{j=1}^{j=3} n_j(\bar{x}_j - \bar{\bar{x}})^2 + (n_1 - 1)S_1^2 + (n_2 - 1)S_2^2 + (n_3 - 1)S_3^2$$

$$4386 = 2871 + 1515$$

After the sums of squares and degrees of freedom are found, the rest of the table is easy to fill in. The mean square is the SS value divided by the degrees of freedom. The factor mean square is $2871/2 = 1435$ and the error mean square is $1515/12 = 126$. The test statistic is the factor mean square divided by the error mean square or $1435/126 = 11.37$. This test statistic has an

F-distribution. The p-value is the area in the upper tail of this distribution beyond 11.37.

The test H_0: $\mu_A = \mu_B = \mu_C$ versus H_a: at least two of the means are different is based on the statistic $F = 1435/126 = 11.37$. This test statistic has an F-distribution with $k-1$ and $n-k$ degrees of freedom. The p-value corresponding to this computed value of F is 0.002. The null hypothesis would be rejected at any alpha value greater than 0.002. The above discussion is generalized in Table 2-9. The basic algebraic result is SS(Total) = SST + SSE.

Table 2-9

Source of variation	Degrees of freedom	Sum of squares	Mean squares	F-statistic
Treatments	$k-1$	SST	$MST = SST/(k-1)$	$F = MST/MSE$
Error	$n-k$	SSE	$MSE = SSE/(n-k)$	
Total	$n-1$	SS(Total)		

The data structure for the above analysis required the data to be placed into three separate columns. Another data structure that is often encountered in statistical packages is shown in Fig. 2-3. This form is called stacked. The Minitab pull-down **Stat** $\Rightarrow$ **ANOVA** $\Rightarrow$ **Oneway** is used when the data are in this form.

Figure 2-4 gives the dialog box produced by the pull-down **Stat** $\Rightarrow$ **ANOVA** $\Rightarrow$ **Oneway**. When this form of the completely randomized design analysis is used, multiple comparisons of means may be requested. This topic will be discussed in Section 2-5. Some additional graphical output for this pull-down is shown in Figs. 2-5 and 2-6. Both of these graphics show that the average distance for brands A and B are close. Brand C produces distances 25 to 30 feet longer than brands A and B. This accounts for the significant p-value ($p = 0.002$).

EXAMPLE 2-2
Use Excel to analyze the data in Example 2-1.

SOLUTION
When using Excel to do the analysis, put the data along with labels in A1:C6. Use the pull-down sequence **Tools** $\Rightarrow$ **Data Analysis**. From the Data Analysis

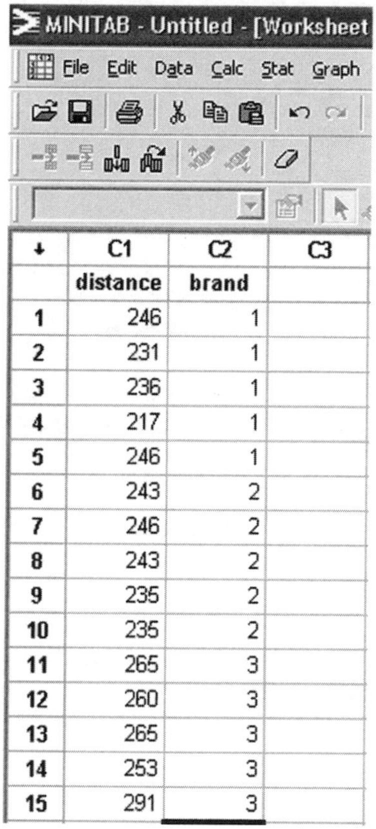

↓	C1	C2	C3
	distance	brand	
1	246	1	
2	231	1	
3	236	1	
4	217	1	
5	246	1	
6	243	2	
7	246	2	
8	243	2	
9	235	2	
10	235	2	
11	265	3	
12	260	3	
13	265	3	
14	253	3	
15	291	3	

MINITAB - Untitled - [Worksheet

File Edit Data Calc Stat Graph

Fig. 2-3.

One-Way Analysis of Variance

Response: distance

Factor: brand

☐ Store residuals
☐ Store fits

Confidence level: 95.0

Select   Comparisons...   Graphs...

Help   OK   Cancel

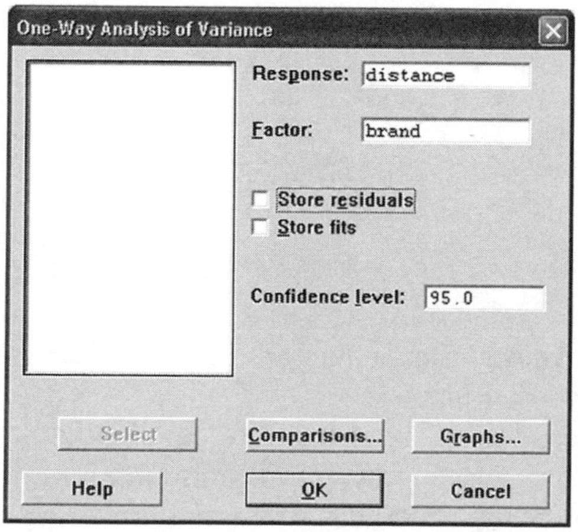

Fig. 2-4.

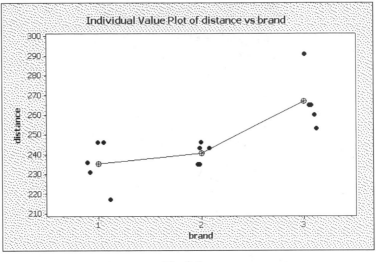

Fig. 2-5.

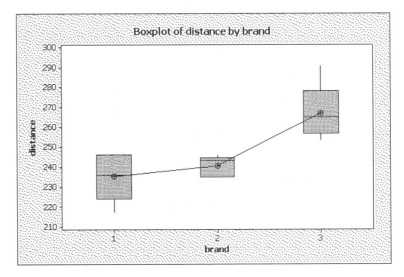

Fig. 2-6.

dialog box, select **Anova: Single Factor**. Fill out the Anova: Single Factor dialog box as shown in Fig. 2-7.

The 0.05 critical F-value is shown to equal 3.88529 in Fig. 2-8. The calculated F-value exceeds this by a considerable amount. This would allow the classical method as well as the p-value method to be used to perform the test. The output produced by Excel is the same as that produced by Minitab.

	A	B	C	D	E	F	G	H	I	J
1	Brand A	Brand B	Brand C	**Anova: Single Factor**					? ✕	
2	246	243	265	Input						
3	231	246	260	Input Range:		$A$1:$C$6			OK	
4	236	243	265	Grouped By:		⦿ Columns			Cancel	
5	217	235	253			○ Rows			Help	
6	246	235	291	☑ Labels in First Row						
7				Alpha: 0.05						
8										
9				Output options						
10				⦿ Output Range:		$D$1				
11				○ New Worksheet Ply:						
12				○ New Workbook						
13										
14										
15										

Fig. 2-7.

Microsoft Excel - Book1

File Edit View Insert Format Tools Data Window Help

Arial   ▼ 10 ▼ **B** *I* U

	A	B	C	D	E	F	G	H	I	J	K
1	Brand A	Brand B	Brand C	Anova: Single Factor							
2	246	243	265								
3	231	246	260	SUMMARY							
4	236	243	265	*Groups*	*Count*	*Sum*	*Average*	*Variance*			
5	217	235	253	Brand A	5	1176	235.2	145.7			
6	246	235	291	Brand B	5	1202	240.4	25.8			
7				Brand C	5	1334	266.8	207.2			
8											
9											
10				ANOVA							
11				*Source of Variation*	*SS*	*df*	*MS*	*F*	*P-value*	*F crit*	
12				Between Groups	2870.933	2	1435.467	11.37153	0.001698	3.88529	
13				Within Groups	1514.8	12	126.2333				
14											
15				Total	4385.733	14					

Fig. 2-8.

Table 2-10   Number of e-mails sent by different age groups.

20 or less	20 < age ≤ 40	40 < age ≤ 60	Above 60
41	44	43	34
43	40	40	37
45	37	42	37
44	43	41	41
42	41	41	37
48	43	39	39
49	42	45	36
48	40	38	41
47	43	40	42
45	42	45	37
47	44	34	37
45	42	43	33
46	40	43	42
45	37	42	36
48		42	
40		39	
44			
39			

EXAMPLE 2-3

Now that the analysis of the completely randomized design has been discussed, some additional examples will help shed some light on the analysis. Four age groups of Internet users have been polled, with the results shown in Table 2-10. The response variable is the number of e-mails sent per week. The output for this example is as follows.

SOLUTION

One-way ANOVA: Age1, Age2, Age3, Age4
Analysis of Variance

```
Source    DF    SS                 MS       F          P
Factor    3     390.09    130.03   17.57    0.000
Error     58    429.26      7.40
Total     61    819.35
                                   Individual 95% CIs For Mean
                                   Based on Pooled StDev
Level     N     Mean      StDev    - - - - -+- - - - -+- - - - -+- - - -
Age1      18    44.778    2.901                         (- -*- -)
Age2      14    41.286    2.268              (- -*- -)
Age3      16    41.063    2.768              (- -*- -)
Age4      14    37.786    2.833    (- -*- -)
                                   - - - - +- - - - - -+- - - -+- - - -
Pooled StDev=  2.720                 39.0      42.0      45.0
```

Figure 2-9 gives a dot plot for the four samples. The mean for each sample is shown by a line within the dot plot. The means for the four samples are statistically different.

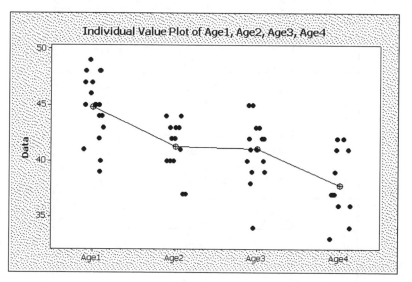

Fig. 2-9.

EXAMPLE 2-4

Table 2-11 gives a different set of data. The data in this table is more variable than the data in Table 2-10 (Example 2-3). The output for this example with the new data is as follows. Figure 2-10 is a dot plot of the data in Table 2-11.

Table 2-11   Number of e-mails sent by different age groups (a second set of data).

20 or less	20 < age ≤ 40	40 < age ≤ 60	Above 60
64	64	55	39
28	28	43	38
33	33	53	45
83	83	18	44
37	37	59	29
31	31	32	30
60	60	27	50
40	40	30	63
56	56	38	29
45	45	25	31
10	10	42	14
26	26	45	48
49	49	20	52
25	25	46	38
52		46	
62		36	
31			
25			

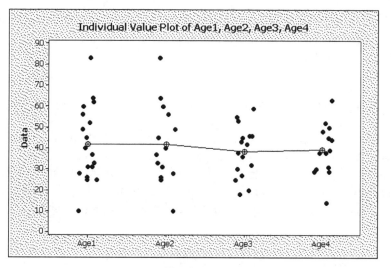

Fig. 2-10.

SOLUTION

One-way ANOVA: Age1, Age2, Age3, Age4
Analysis of Variance

Source	DF	SS	MS	F	P
Factor	3	161	54	0.21	0.888
Error	58	14655	253		
Total	61	14815			

Individual 95% CIs For Mean
Based on Pooled StDev

Level	N	Mean	StDev	- - - -+- - - -+- - - -+- - - -
Age1	18	42.06	18.23	(- - - - -*- - - - -)
Age2	14	41.93	19.09	(- - - - -*- - - - -)
Age3	16	38.44	12.35	(- - - - -*- - - - -)
Age4	14	39.29	12.34	(- - - - -*- - - - -)

```
                               - - - -+- - - -+- - - -+- - - -
Pooled StDev=      15.90        36.0        42.0       48.0
```

In the first example, the variation within the samples is relatively small. Note that the four sample standard deviations are 2.901, 2.268, 2.768, and 2.833. They are small when compared with the sample standard deviations in the second example: 18.23, 19.09, 12.35, and 12.34. This causes the error sums of squares to increase from 819.35 in the first example to 14,655 in the second example. This increase in the error sums of squares causes the *p*-value to go from 0.000 to 0.888; i.e., we go from inferring that there are highly significant differences in means to inferring there are no differences in means.

EXAMPLE 2-5

Finally, consider the following example that compares the blood pressure changes of patients with high blood pressure. The patients are randomly divided into three groups. One group is treated with a diet that is very restrictive, another group is treated with a strict exercise program, and the third serves as a control group. The response variable is the change in diastolic blood pressure after six months of treatment. In this highly unlikely example, there is no variation within the three groups (Table 2-12).

Table 2-12   An experiment with no variation within treatments.

Diet	Exercise	Control
10	13	2
10	13	2
10	13	2
10	13	2
10	13	2

The error term in the analysis of variance has zero sums of squares. All the variation is between the three treatments. There is no variation within treatments.

SOLUTION

One-way ANOVA: Diet, Exercise, Control

```
Analysis of Variance
Source    DF    SS           MS            F      P
Factor     2    323.3333     161.6667      *      *
Error     12      0.0000       0.0000
Total     14    323.3333
```

```
                                    Individual 95% CIs For Mean
                                    Based on Pooled StDev
Level        N     Mean     StDev   - - -+ - - - + - - - + - - - + - -
Diet         5   10.0000   0.0000                              *
Exercise     5   13.0000   0.0000                                    *
Control      5    2.0000   0.0000   *

                                    - - -+ - - - + - - - + - - - + - -
Pooled StDev=       0.0000          3.0     6.0     9.0    12.0
```

In this case, all the variation is due to differences between the treatment groups. The error term is zero. In other words, there is no variation within the treatment groups. The three population means would be declared different (Fig. 2-11).

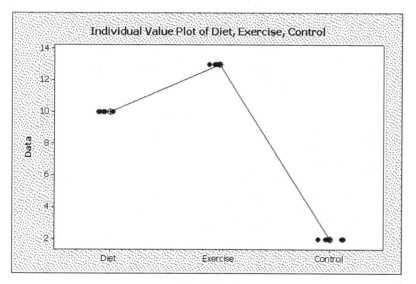

Fig. 2-11.

2-3 The Randomized Complete Block Design

Purpose of the test: The purpose of the test is to compare the means of several treatments when they have been administered in blocks. The treatments have been randomly assigned to experimental units within the blocks.

Assumptions: (1) The probability distributions of observations corresponding to all block–treatment combinations are normal. (2) The variances of all probability distributions are equal.

EXAMPLE 2-6

We are interested in comparing the distances that three different brands of balls travel, when hit by the club known as the driver. Five golfers of varying ability each hit the three brands in random order. The letters C, B, and A are randomly pulled out of a hat in that order. Jones will hit brand C, followed by brand B, followed by brand A. Similarly, the letters A, C, and B are randomly chosen, so that Smith will hit brand A, followed by brand C, followed by brand B. Continuing in this manner will ensure the random assignment of treatments within blocks. The three brands of balls are the treatments and the five golfers are the blocks.

The distance that each ball travels is the response variable. These distances are given in Table 2-13 for each of the five golfers.

Table 2-13   Statistical layout showing treatments and blocks.

Block	Block	Treatments		
Totals (mean)	Golfer	Brand A	Brand B	Brand C
767 (256)	Jones	250	255	262
690 (230)	Smith	225	235	230
837 (279)	Long	270	282	285
721 (240)	Carroll	235	240	246
658 (219)	Reed	215	220	223

SOLUTION

The data is entered into the Minitab worksheet as shown in Fig. 2-12.

The pull-down **Stat** $\Rightarrow$ **ANOVA** $\Rightarrow$ **Two Way** gives the dialog box shown in Fig. 2-13, which is filled in as shown. The output shown below is given by Minitab.

Two-way ANOVA: Distance versus Block, Treatment

```
Analysis of Variance for Distance
Source        DF     SS        MS        F         P
Block          4   6525.73   1631.43   203.08    0.000
Treatment      2    277.73    138.87    17.29    0.001
Error          8     64.27      8.03
Total         14   6867.73
```

↓	C1	C2	C3	C4
	Distance	Block	Treatment	
1	250	1	1	
2	225	2	1	
3	270	3	1	
4	235	4	1	
5	215	5	1	
6	255	1	2	
7	235	2	2	
8	282	3	2	
9	240	4	2	
10	220	5	2	
11	262	1	3	
12	230	2	3	
13	285	3	3	
14	246	4	3	
15	223	5	3	

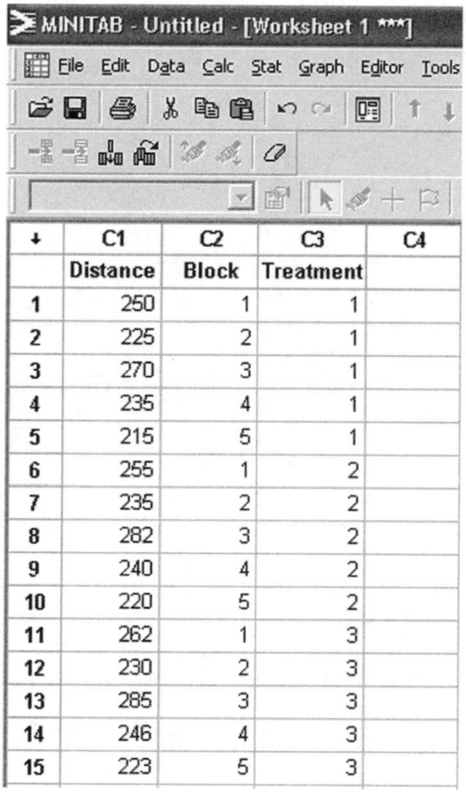

Fig. 2-12.

```
                  Individual 95% CI
Block     Mean    - - - -+- - - - - - -+- - - - - +- - - - - - -+- - - -
1         255.7                                  (-*-)
2         230.0            (-*-)
3         279.0                                                (-*-)
4         240.3                      (-*-)
5         219.3     (-*-)
                  - - - -+- - - - -+- - - - - -+- - - - -+- - - - - - -
                    220.0     240.0       260.0     280.0
```

```
                  Individual 95% CI
Treatment   Mean  - - - -+- - - - - -+- - - - - -+- - - - - -+- - -
1          239.0  (- - - * - - -)
2          246.4                 (- - - * - - -)
3          249.2                             (- - - * - - -)
                  - - - -+- - - - - -+- - - - - -+- - - - - -+- - -
                    240.0      244.0      248.0      252.0
```

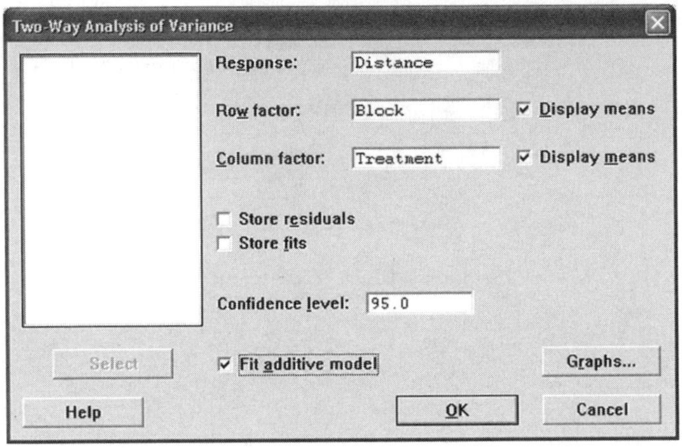

Fig. 2-13.

In the two-way ANOVA output above, the following equations hold. The total sums of squares is

$$\text{total sums of squares} = \sum (x - \bar{\bar{x}})^2 = 6867.73$$

In a block design this total variation is expressed as a sum of three sources as follows:

$$\sum (x - \bar{\bar{x}})^2 = \sum_{j=1}^{k} b(\bar{x}[T]_j - \bar{\bar{x}})^2 + \sum_{i=1}^{b} k(\bar{x}[B]_i - \bar{\bar{x}})^2$$
$$+ \sum_{j=1}^{k} \sum_{i=1}^{b} (x_{ij} - \bar{x}[T]_j - \bar{x}[B]_i + \bar{\bar{x}})^2$$

where:
  $\bar{\bar{x}}$ is the grand mean over all observations
  $\bar{x}[T]_j$ is the mean of the observations in the jth treatment
  $\bar{x}[B]_i$ is the mean of the observations in the ith block
  k is the number of treatments
  b is the number of blocks

or, in simpler terms,

$$\text{SS(total)} = \text{SST} + \text{SSB} + \text{SSE}$$

or, referring to the output,

$$6867.73 = 277.73 + 6525.73 + 64.27$$

The general form of the block ANOVA can be written as given in Table 2-14.

Table 2-14

Source of variation	Degrees of freedom	Sum of squares	Mean squares	F-statistic
Treatments	$k - 1$	SST	$MST = SST/(k-1)$	$F = (MST/MSE)$
Blocks	$b - 1$	SSB	$MSB = SSB/(b-1)$	$F = (MSB/MSE)$
Error	$n - k - b + 1$	SSE	$MSE = SSE/(n-k-b+1)$	
Total	$n - 1$	SS(Total)		

Referring to the output for this example, we see that most of the variation is caused by the difference in the five blocks or the five golfers. The p-values lead us to believe that there are differences in golfers as well as brands of balls. The p-values for blocks allows us to test the hypothesis

$$H_0: \mu_{\text{block1}} = \cdots = \mu_{\text{block5}} \text{ versus}$$
$$H_a: \text{at least some two of the block means differ}$$

The p-values for treatments allows us to test the hypothesis

$$H_0: \mu_A = \mu_B = \mu_C \text{ versus}$$
$$H_a: \text{at least two of the treatment means are different}$$

The confidence intervals for blocks allow us to compare golfers, and the confidence intervals for treatments allow us to compare brands. The confidence intervals for brands suggest that brand A gives smaller distances on the average than does brand B or C. They also suggest that there may be no difference between B and C.

EXAMPLE 2-7
Consider the same experiment with the data in Table 2-15.

SOLUTION
The output for this experiment is as follows.

Table 2-15

Block	Block	Treatments		
Totals (mean)	Golfer	Brand A	Brand B	Brand C
767 (256)	Jones	250	255	262
767 (256)	Smith	255	250	262
767 (256)	Long	245	257	265
767 (256)	Carroll	250	255	262
767 (256)	Reed	250	255	262

Two-way ANOVA: Distance versus Block, Treatment

```
Analysis of Variance for Distance

Source        DF       SS        MS        F        P
Block          4      0.0       0.0      0.00    1.000
Treatment      2    408.9     204.5     19.38    0.001
Error          8     84.4      10.5
Total         14    493.3

                     Individual 95% CI
Block   Mean    - - - - + - - - - - + - - - - - + - - - - - - + - - -
1       255.7   ( - - - - - - - - - - - * - - - - - - - - - - - )
2       255.7   ( - - - - - - - - - - - * - - - - - - - - - - - )
3       255.7   ( - - - - - - - - - - - * - - - - - - - - - - - )
4       255.7   ( - - - - - - - - - - - * - - - - - - - - - - - )
5       255.7   ( - - - - - - - - - - - * - - - - - - - - - - - )
                - - - - + - - - - - - + - - - - - + - - - - - + - - -
                   252.5       255.0       257.5       260.0

                     Individual 95% CI
Treatment Mean  - - - - + - - - - - - + - - - - - + - - - - - + - - -
1       250.0  ( - - - * - - - )
2       254.4          ( - - - * - - - )
3       262.6                           ( - - - * - - - )
                - - - - + - - - - - - + - - - - - + - - - - - + - - -
                   250.0       255.0       260.0       265.0
```

In this example, the golfers were of comparable ability and blocking on golfers was not effective in reducing variation. The experiment should be designed as a completely randomized experiment if the golfers are of comparable ability.

EXAMPLE 2-8
Analyze the block design using Excel.

SOLUTION
The Excel analysis is as follows. Enter the data into the work sheet and execute the pull-down sequence **Tools ⇒ Data Analysis.** From the Data Analysis dialog box choose **ANOVA: Two-Factor Without Replication** as given in Fig. 2-14.

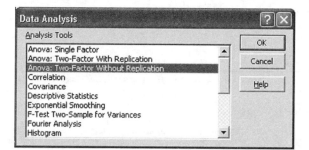

Fig. 2-14.

Fill in the ANOVA: Two-Factor Without Replication dialog box as shown in Fig. 2-15. The output is given in Fig. 2-16.

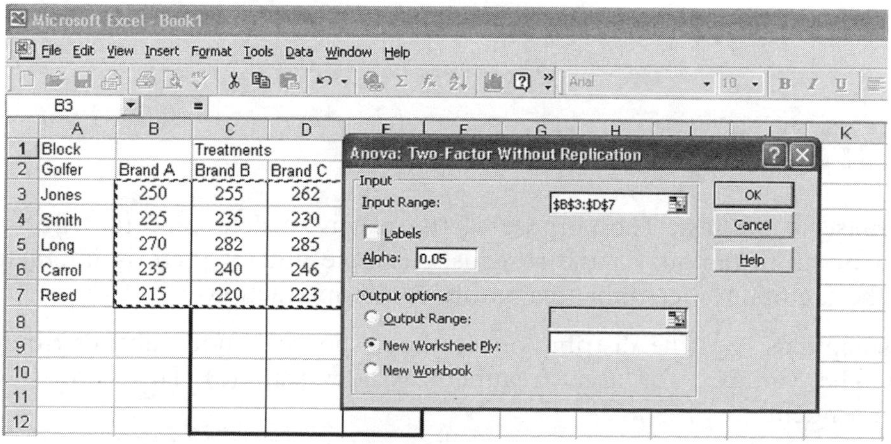

Fig. 2-15.

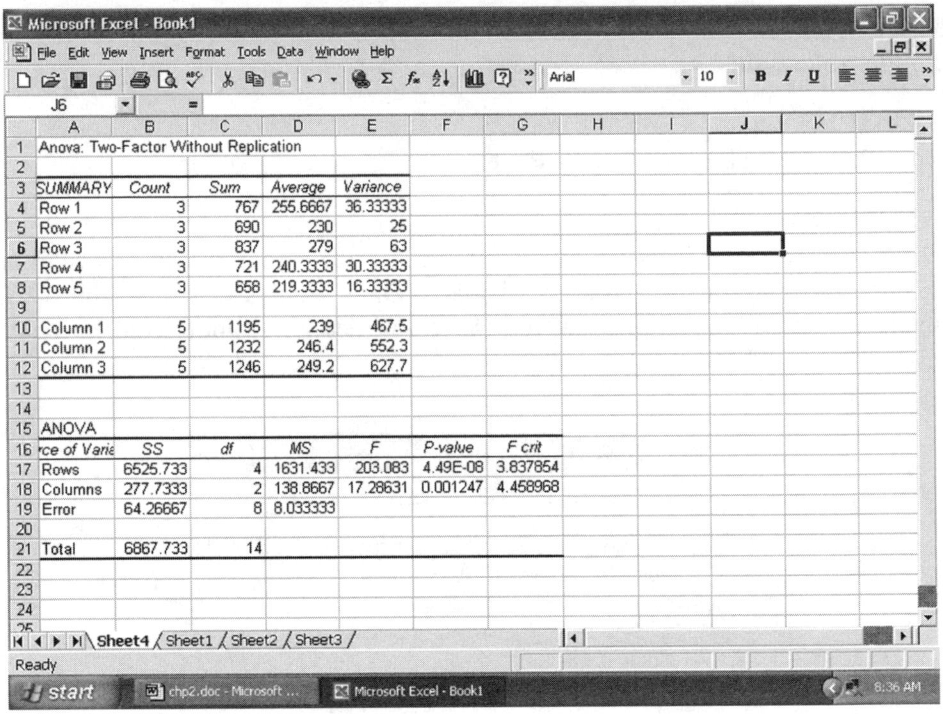

	A	B	C	D	E	F	G
1	Anova: Two-Factor Without Replication						
2							
3	SUMMARY	Count	Sum	Average	Variance		
4	Row 1	3	767	255.6667	36.33333		
5	Row 2	3	690	230	25		
6	Row 3	3	837	279	63		
7	Row 4	3	721	240.3333	30.33333		
8	Row 5	3	658	219.3333	16.33333		
9							
10	Column 1	5	1195	239	467.5		
11	Column 2	5	1232	246.4	552.3		
12	Column 3	5	1246	249.2	627.7		
13							
14							
15	ANOVA						
16	rce of Variă	SS	df	MS	F	P-value	F crit
17	Rows	6525.733	4	1631.433	203.083	4.49E-08	3.837854
18	Columns	277.7333	2	138.8667	17.28631	0.001247	4.458968
19	Error	64.26667	8	8.033333			
20							
21	Total	6867.733	14				

Fig. 2-16.

The Excel output is the same as the Minitab output. The only thing that Excel gives that Minitab does not is the critical F-value for blocks, 3.84, and the critical F-value for treatments, 4.46. You don't really need these values if you use the p-value approach to testing.

2-4 Factorial Experiments

Purpose of the test: The purpose of the test is to determine the effects of two or more factors on the response variable and, if there is interaction of the factors, to determine the nature of the interaction.

Assumptions: (1) The distribution of the response is normally distributed. (2) The variance for each treatment is identical. (3) The samples are independent.

EXAMPLE 2-9

Suppose we wished to consider the effect of two factors on blood pressure. *Factor A* is diabetes. The two *levels* are "present" and "absent." *Factor B* is weight. The two *levels* are "overweight" and "normal." Five diabetics of normal weight, five diabetics who are overweight, five non-diabetics of normal weight, and five non-diabetics who are overweight are randomly selected. None of the twenty were on medication for high blood pressure. The diastolic blood pressure of the twenty participants is measured and the results are given in Table 2-16. We are interested in the *interaction* of weight and diabetes

Table 2-16

	Normal weight	Overweight
Non-diabetic	75, 80, 83, 85, 65	85, 80, 90, 95, 88
Diabetic	85, 90, 95, 90, 86	90, 95, 100, 105, 110

on the blood pressure. If there is no significant interaction, then we are interested in the effect of diabetes on blood pressure and in the effect of weight on blood pressure. We refer to this as a *2 by 2-factorial experiment* with 5 *replicates* whose response variable is diastolic blood pressure.

SOLUTION

The data in Table 2-16 is entered into the Minitab worksheet as shown in Fig. 2-17. The pull-down **Stat** $\Rightarrow$ **ANOVA** $\Rightarrow$ **Balanced Anova** gives the Balanced Analysis of Variance dialog box in Fig. 2-18. Diastolic is entered as the response variable and, for the model, Factor A, Factor B, Factor A * Factor B (this term represents the interaction) is entered.

The total sums of squares are expressed as a sum of Factor A sums of squares, Factor B sums of squares, interaction sums of squares, and error sums of squares by the following expression:

$$rb\sum_{i=1}^{a}(\bar{x}[A]_i - \bar{\bar{x}})^2 + ra\sum_{j=1}^{b}(\bar{x}[B]_j - \bar{\bar{x}})^2 + r\sum_{i=1}^{a}\sum_{j=1}^{b}(\bar{x}[AB]_{ij} - \bar{x}[A]_i - \bar{x}[B]_j + \bar{\bar{x}})^2$$

$$+ \sum_{i=1}^{a}\sum_{j=1}^{b}\sum_{k=1}^{r}(x_{ijk} - \bar{x}[AB]_{ij})^2$$

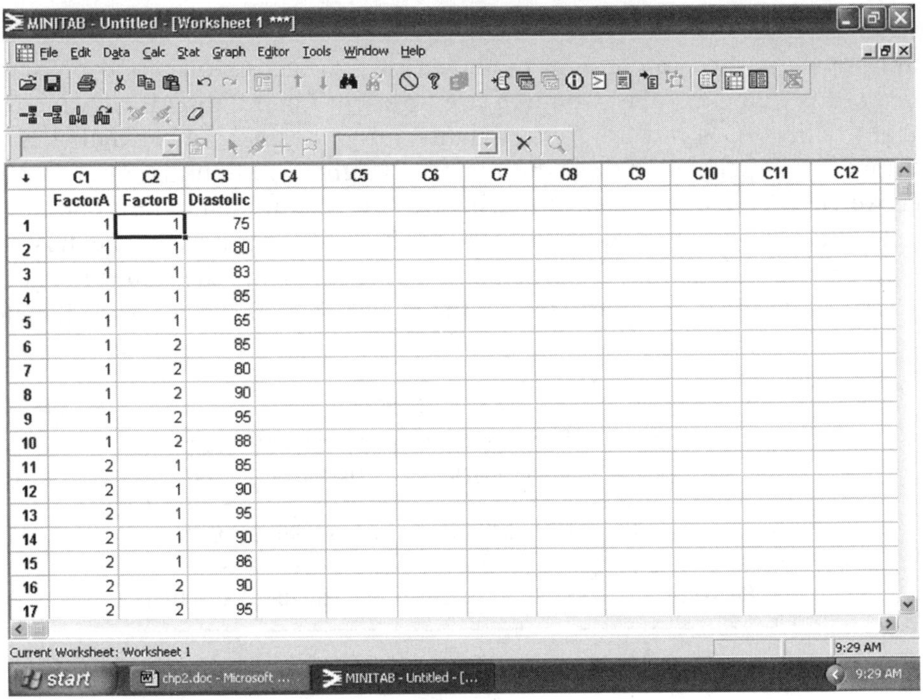

Fig. 2-17.

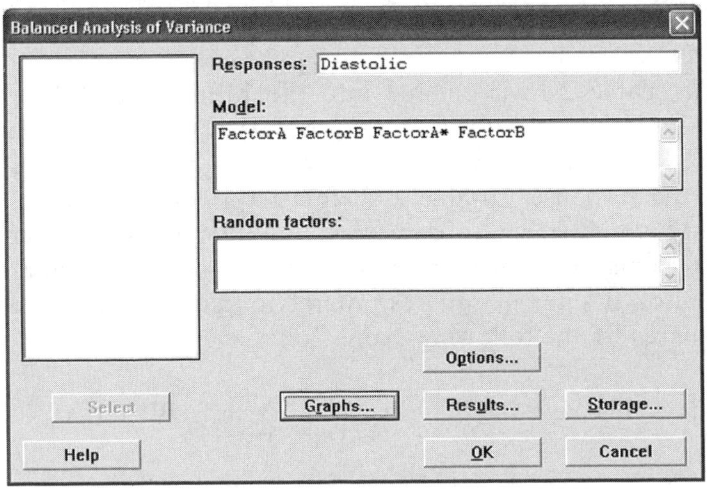

Fig. 2-18.

where

$\bar{x}[AB]_{ij}$ is the mean of the response in the ijth treatment (mean of the treatment when the factor A level is i and the factor B level is j)
$\bar{x}[A]_i$ is the mean of the responses when the factor A level is i
$\bar{x}[B]_j$ is the mean of the responses when the factor B level is j
$\bar{\bar{x}}$ is the mean of all responses
a is the number of factor A levels, b is the number of factor B levels, c is the number of replicates

or, in simpler terms,

$$SS(total) = SSA + SSB + SSAB + SSE$$

The general form of the two-factor factorial is as shown in Table 2-17.

Table 2-17

Source of variation	Degrees of freedom	Sum of squares	Mean squares	F-statistic
Factor A	$a-1$	SSA	$MSA = SSA/a-1$	$F = MSA/MSE$
Factor B	$b-1$	SSB	$MSB = SSB/b-1$	$F = MSB/MSE$
Interaction	$(a-1)(b-1)$	SSAB	$MSAB = SSAB/(a-1)(b-1)$	$F = MSAB/MSE$
Error	$n-ab$	SSE	$MSE = SSE/n-ab$	
Total	$n-1$	SS(total)		

The output is as follows.

ANOVA: Diastolic versus FactorA, FactorB

```
Factor     Type      Levels   Values

FactorA    fixed     2          1        2
FactorB    fixed     2          1        2

Analysis of Variance for Diastoli

Source     DF     SS        MS        F       P
FactorA    1      720.00    720.00    16.62   0.001
FactorB    1      540.80    540.80    12.48   0.003
```

```
FactorA * Factor B    1        00.80   0.80    0.02     0.894
Error                16       693.20  43.33
Total                19      1954.80
```

Means

```
FactorA    N        Diastoli
1         10        82.600
2         10        94.600

FactorB    N        Diastoli
1         10        83.400
2         10        93.800
```

Note first the partitioning of the total sums of squares:

$$1954.80 = 720.00 + 540.80 + 0.80 + 693.20$$

The interaction term is FactorA * FactorB and is non-significant, as is indicated by the p-values $= 0.894$.

The interaction plot is obtained by **Stat $\Rightarrow$ ANOVA $\Rightarrow$ Interactions Plot**. The dialog box is shown in Fig. 2-19. The interaction plot is shown in Fig. 2-20.

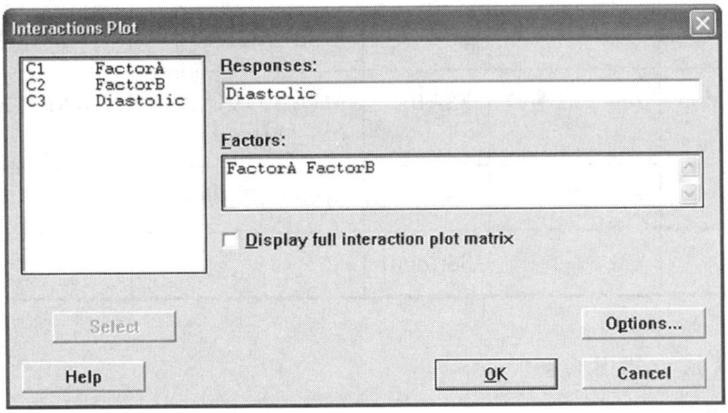

Fig. 2-19.

In Fig. 2-20 the solid line shows the response for non-diabetics and the dotted line shows the response for diabetics. The response for both diabetics and non-diabetics shows an increase in diastolic blood pressure when the weight level changes from normal weight to overweight. The fact that the lines are nearly parallel indicates there is no interaction. The other interaction plot is obtained when B is entered first and A second in the Factors part of

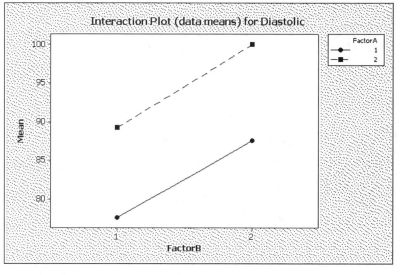

Fig. 2-20.

the interactions dialog box. The dialog box and interaction plot are shown in Figs. 2-21 and 2-22 respectively.

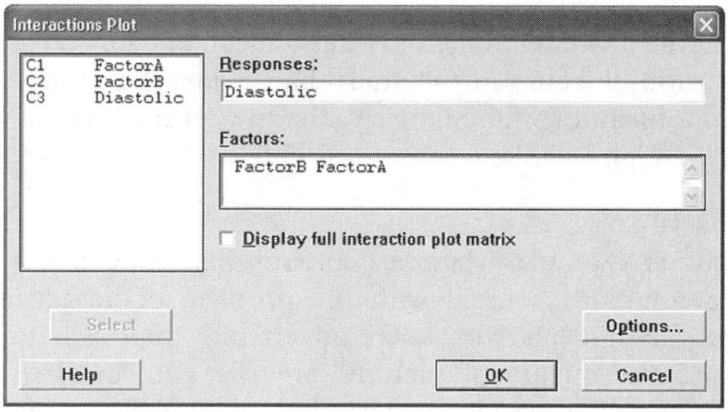

Fig. 2-21.

Now, turning to the main effects, we see that both main effects are significant; that is, the *p*-values are much smaller than 0.05.

```
Means
FactorA    N    Diastoli
1         10    82.600
2         10    94.600
```

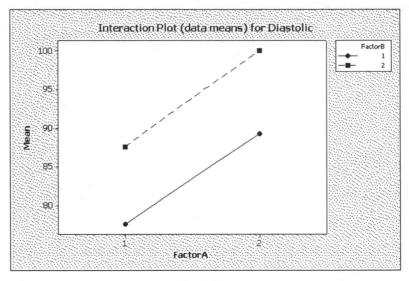

Fig. 2-22.

FactorB	N	Diastoli
1	10	83.400
2	10	93.800

Notice that the low level of factor A has a mean of 82.6 and a high level of 94.6; that is, non-diabetics in the study had a mean diastolic of 82.6 and diabetics had a mean of 94.6. Similarly, the low level of weight had a mean of 83.4 and the high level had a mean of 93.8.

EXAMPLE 2-10
Consider another example, where a company has developed a new digital camera. The company is faced with the problem of advertising the new camera. One factor deals with what advertising approach to emphasize. The price and the quality of pictures are the two levels of advertising approach the company decides to use. The other factor of interest is the advertising medium to use. The levels of advertising medium that the company will use are radio, newspaper, and Internet. The response variable is the number of weekly sales. The data are shown in Table 2-18.

SOLUTION
The Minitab output is as follows. We notice first that the interaction is significant. Thus our objective is to explain the nature of the interaction. In doing this we will discover what the experiment has really found about what, and how sales are affected.

Table 2-18 Sales as affected by advertising medium and advertising approach.

Factor B, advertising approach	Factor A, advertising medium		
	Radio	Newspaper	Internet
Price	15, 20, 17	30, 32, 35	25, 28, 22
Quality	17, 20, 13	25, 27, 22	35, 37, 40

ANOVA: Sales versus FactorA, FactorB

```
Factor      Type      Levels   Values
FactorA     fixed     3        1       2       3
FactorB     fixed     2        1       2

Analysis of Variance for Sales

Source                 DF    SS        MS        F        P
FactorA                 2    680.11    340.06    43.72    0.000
FactorB                 1      8.00      8.00     1.03    0.331
FactorA * FactorB       2    309.00    154.50    19.86    0.000
Error                  12     93.33      7.78
Total                  17   1090.44

Means

FactorA     N     Sales
1           6     17.000
2           6     28.500
3           6     31.167
FactorB     N     Sales
1           9     24.889
2           9     26.222
```

The first thing we should do is to plot the interaction graphs. Look at the results from all angles. The interaction plots are given in Figs. 2-23 and 2-24. In Fig. 2-23, the solid line describes radio sales, the dotted line describes Internet sales, and the dashed line describes newspaper sales. Radio sales are relatively low and are the same for both the price and the quality approach. For newspaper advertising sales are higher for the price approach than for the quality approach. The sales are greater for quality approach than for price approach for Internet advertising. The greatest sales are for Internet advertising where the quality approach is used.

In Fig. 2-24, the solid line is for the price approach to advertising the digital camera and the dashed line is for the quality approach. When the

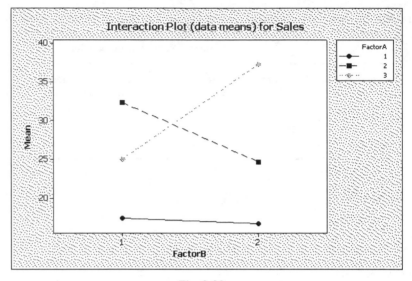

Fig. 2-23.

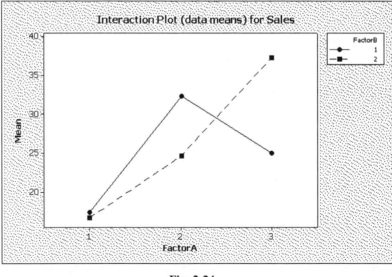

Fig. 2-24.

price approach is used, the greatest sales are found when advertising in newspapers. When the quality approach is used, the Internet approach to advertising is the best. These two interaction plots say the same things but look at the statistics from two different directions.

EXAMPLE 2-11
Work Example 2-10 using Excel.

SOLUTION
The Excel analysis for the data will now be illustrated. First, the data is entered into the worksheet as shown in Fig. 2-25.

	A	B	C	D	E
1		Radio	Newspaper	Internet	
2	Price	15	30	25	
3	Price	20	32	28	
4	Price	17	35	22	
5	Quality	17	25	35	
6	Quality	20	27	37	
7	Quality	13	22	40	
8					

Fig. 2-25.

The pull-down **Tools** $\Rightarrow$ **Data Analysis** gives the Data Analysis dialog box shown in Fig. 2-26. Choose Anova: Two-Factor With Replication.

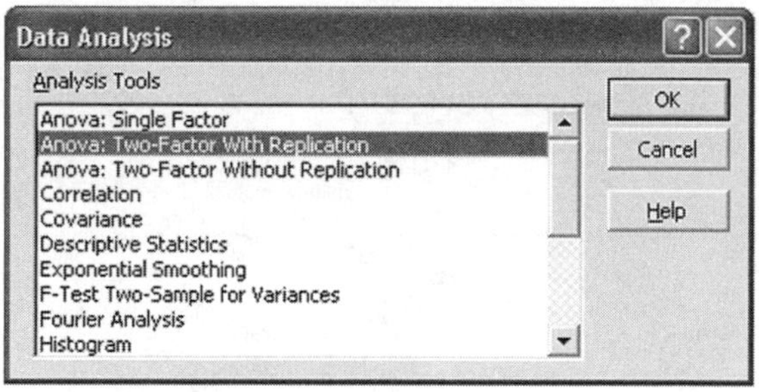

Fig. 2-26.

Look at Fig. 2-25 and fill in the dialog box as shown in Fig. 2-27. The following Excel output, shown in Figs. 2-28 and 2-29, is generated.

The output in Fig. 2-29 may be compared with the Minitab output in order to identify the various parts of the Excel Anova.

Anova: Two-Factor With Replication

Input	
Input Range:	$A$1:$D$7
Rows per sample:	3
Alpha:	0.05

Output options
- Output Range:
- New Worksheet Ply:
- New Workbook

OK
Cancel
Help

Fig. 2-27.

	A	B	C	D	E	F
1	Anova: Two-Factor With Replication					
2						
3	SUMMARY	Radio	Newspape	Internet	Total	
4	*Price*					
5	Count	3	3	3	9	
6	Sum	52	97	75	224	
7	Average	17.33333	32.33333	25	24.88889	
8	Variance	6.333333	6.333333	9	47.61111	
9						
10	*Quality*					
11	Count	3	3	3	9	
12	Sum	50	74	112	236	
13	Average	16.66667	24.66667	37.33333	26.22222	
14	Variance	12.33333	6.333333	6.333333	87.69444	
15						
16	*Total*					
17	Count	6	6	6		
18	Sum	102	171	187		
19	Average	17	28.5	31.16667		
20	Variance	7.6	22.7	51.76667		
21						

Fig. 2-28.

EXAMPLE 2-12

Factorial designs with more than two factors are common in statistics. For example, suppose we wanted to investigate the effect of three factors on the amount of dirt removed from a standard load of clothes. The three factors are brand of laundry detergent, A, water temperature, B, and type of detergent, C. The two levels of brand of detergent are brand X and brand Y. The two levels of water temperature are warm and hot. The two levels of

22							
23	ANOVA						
24	rce of Varia	SS	df	MS	F	P-value	F crit
25	Sample	8	1	8	1.028571	0.330507	4.747221
26	Columns	680.1111	2	340.0556	43.72143	3.09E-06	3.88529
27	Interaction	309	2	154.5	19.86429	0.000156	3.88529
28	Within	93.33333	12	7.777778			
29							
30	Total	1090.444	17				

Fig. 2-29.

detergent type are powder and liquid. The factorial design that applies to this experiment is called a 2^3 factorial design. There are eight treatments possible in a 2^3 design. They are shown in Table 2-19.

Table 2-19 Treatments are composed of different combinations of factor levels.

Treatment	Detergent	Water temp.	Detergent type
1	X	Warm	Powder
2	X	Warm	Liquid
3	X	Hot	Powder
4	X	Hot	Liquid
5	Y	Warm	Powder
6	Y	Warm	Liquid
7	Y	Hot	Powder
8	Y	Hot	Liquid

Suppose this experiment was run with 2 replications. (This would require 16 standard loads of clothes.) The ANOVA for such an experiment is shown in Table 2-20. The expressions for the sums of squares are omitted.

If any of the four interactions are significant, then the nature of the interactions is investigated with interaction plots. If none of the interactions are significant, then the main effects of the three factors are investigated.

Table 2-20   ANOVA breakdown for a 2^3 factorial experiment (3 factors each at 2 levels).

Source of variation	Degrees of freedom	Sum of squares	Mean squares	F-statistic
A	1	SSA	MSA	$F = $ MSA/MSE
B	1	SSB	MSB	$F = $ MSB/MSE
C	1	SSC	MSC	$F = $ MSC/MSE
AB	1	SSAB	MSAB	$F = $ MSAB/MSE
AC	1	SSAC	MSAC	$F = $ MSAC/MSE
BC	1	SSBC	MSBC	$F = $ MSBC/MSE
ABC	1	SSABC	MSABC	$F = $ MSABC/MSE
Error	8	SSE	MSE	
Total	15			

SOLUTION

Suppose the data from this experiment are as shown in Table 2-21. Eight standard loads were randomly assigned to the eight treatments. This experiment was then replicated so that two observations for each treatment were obtained. The steps to follow when using Minitab are shown in Figs. 2-30 and 2-31. The output is as follows.

ANOVA: Response versus A, B, C

```
Factor   Type    Levels   Values
A        fixed   2        1        2
B        fixed   2        1        2
C        fixed   2        1        2

Analysis of Variance for response

Source   DF     SS          MS          F         P
A        1        1.210       1.210       8.80      0.018
B        1      117.723     117.723     856.16      0.000
C        1       97.022      97.022     705.62      0.000
```

```
Source      DF    SS       MS      F       P
A * B       1     0.302    0.302   2.20    0.176
A * C       1     0.023    0.023   0.16    0.696
B * C       1     0.010    0.010   0.07    0.794
A * B * C   1     0.040    0.040   0.29    0.604
Error       8     1.100    0.137
Total       15    217.430
```

Note that none of the interactions are significant, but all main effects are significant.

Table 2-21 Data for 2^3 experiment.

Detergent	Water temp.	Detergent type	Replicate 1	Replicate 2
X	Warm	Powder	15.3	14.9
X	Warm	Liquid	20.4	20.1
X	Hot	Powder	20.5	20.3
X	Hot	Liquid	25.4	25.1
Y	Warm	Powder	16.0	15.1
Y	Warm	Liquid	20.8	19.9
Y	Hot	Powder	21.3	21.1
Y	Hot	Liquid	26.3	25.9

The data are entered in the worksheet as shown in Fig. 2-30.

The pull-down **Stat** $\Rightarrow$ **ANOVA** $\Rightarrow$ **Balanced Anova** gives the Balanced Analysis of Variance dialog box which is filled out as shown in Fig. 2-31.

Figure 2-32 indicates that no interaction is present, since the lines are nearly parallel in all three graphs.

Figure 2-33 gives graphical descriptions of the main effects. The means at the low and high levels of the factors are as follows.

```
Means

Brand    N    Response
1        8    20.250
2        8    20.800
```

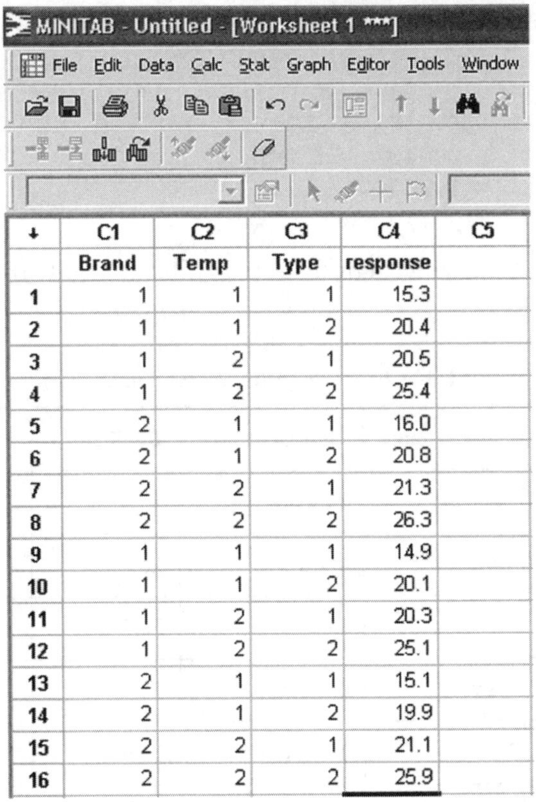

↓	C1 Brand	C2 Temp	C3 Type	C4 response	C5
1	1	1	1	15.3	
2	1	1	2	20.4	
3	1	2	1	20.5	
4	1	2	2	25.4	
5	2	1	1	16.0	
6	2	1	2	20.8	
7	2	2	1	21.3	
8	2	2	2	26.3	
9	1	1	1	14.9	
10	1	1	2	20.1	
11	1	2	1	20.3	
12	1	2	2	25.1	
13	2	1	1	15.1	
14	2	1	2	19.9	
15	2	2	1	21.1	
16	2	2	2	25.9	

Fig. 2-30.

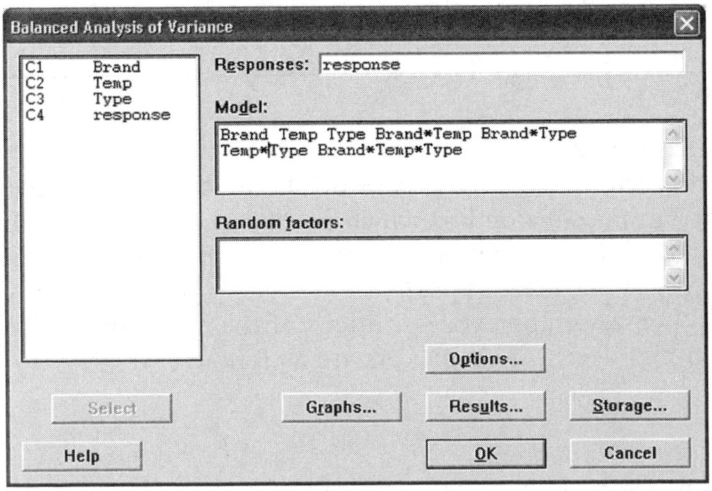

Fig. 2-31.

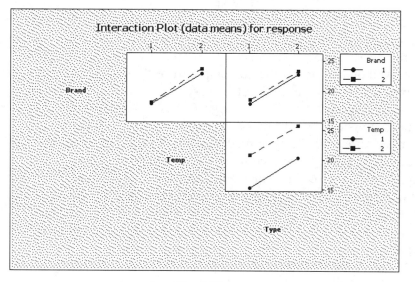

Fig. 2-32.

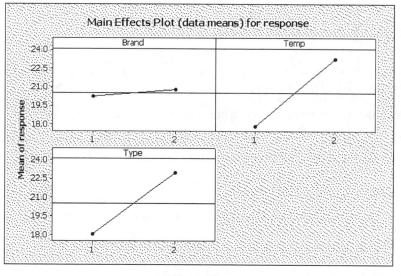

Fig. 2-33.

```
Temp     N     Response
1        8     17.813
2        8     23.238

Type     N     Response
1        8     18.063
2        8     22.988
```

The main effect of brand is $20.800 - 20.250 = 0.55$. That is, Brand Y removes 0.55 more grams of dirt on average than does Brand X. The main effect of temperature is $23.238 - 17.813 = 5.425$. That is, 5.425 more grams of dirt are removed on average at the hot temperature than at the warm temperature of the water. The main effect of detergent type is $22.988 - 18.063 = 4.925$. That is, liquid detergent on average removes 4.925 more grams of dirt than does powder. The brand of detergent (X or Y) is not as important as the temperature and the type of detergent. Using a hot temperature and a liquid detergent would be recommended.

There are no Excel routines for three or more factors, but there are Minitab routines for any number of factors. The number of experimental units required for experiments with a large number of factors becomes very large. For example, a 2^4 factorial experiment with two replications requires 32 experimental units. As the number of factors increases, the interpretation of factorial experiments becomes more difficult. Topics involving large numbers of factors are beyond the scope of this book.

2-5 Multiple Comparisons of Means

Purpose of the test: To compare various combinations of means with combinations of other means.

Assumptions: Vary, depending on the method or procedure used.

The analysis of variance techniques described in the first four sections of this chapter allow you to determine whether there is a difference in a group of means, that is, it allows you to test whether a group of means differs. In the case of testing several means you are testing

$H_0: \mu_A = \mu_B = \cdots = \mu_p$ versus H_a: at least two of the means are different

When the null hypothesis in the analysis of variance is rejected, it is often desirable to know which treatments are responsible for the difference between population means.

To illustrate, suppose an analysis of variance has led to the conclusion that, of four means, not all are equal. Suppose the four sample means are $\bar{x}_1 = 17.1$, $\bar{x}_2 = 28.6$, $\bar{x}_3 = 18.4$, and $\bar{x}_4 = 18.8$. We might be interested in comparing all pairs of means, that is, comparing μ_1 and μ_2, μ_1 and μ_3, μ_1 and μ_4, μ_2 and μ_3, μ_2 and μ_4, and μ_3 and μ_4. Or, we might be interested in the following, for example: (1) Comparing the average of treatments 1, 2, and 3 with the average of treatment 4. (2) Determining whether μ_4 is larger than the other means. (3) Determining whether there is no difference between μ_4, μ_3, and μ_1.

There are many different multiple comparison procedures that deal with these problems. Some of these procedures are as follows: Fisher's least significant difference, Bonferroni's adjustment, Tukey's multiple comparison method, Dunnet's method, Scheffe's general procedure for comparing all possible linear combinations of treatment means, and Duncan's multiple range test. Some require equal sample sizes, while some do not. The choice of a multiple comparison procedure used with an ANOVA will depend on the type of experimental design used and the comparisons of interest to the analyst.

We will illustrate multiple comparison methods using Tukey's multiple comparison method.

EXAMPLE 2-13

An experiment was designed to compare four methods of teaching high school algebra. One method is the traditional chalk-and-blackboard method, referred to as treatment 1. A second method utilizes Excel weekly in the teaching of algebra and is called treatment 2. A third method utilizes the software package Maple weekly and is called treatment 3. A fourth method utilizes both Maple and Excel weekly and is called treatment 4. Sixty students are randomly divided into four groups and the experiment is carried out over a one semester time period. The response variable is the score made on a common comprehensive final in the course. The scores made on the final are shown in Table 2-22. Use Tukey's method to compare all means.

SOLUTION

Enter the data in unstacked form in columns C1, C2, C3 and C4. The pull-down **Stat** $\Rightarrow$ **ANOVA** $\Rightarrow$ **Oneway (Unstacked)** gives the dialog box shown in Fig. 2-34. Fill in the dialog box as shown. Click comparisons. This brings up a new dialog box, shown in Fig. 2-35. Fill in the One-way Multiple Comparisons dialog box as shown. The output below is produced.

One-way ANOVA: Score versus Method

Source	DF	SS	MS	F	P
Method	3	694.6	231.5	7.41	0.000
Error	56	1750.4	31.3		
Total	59	2445.0			

S = 5.591   R-Sq = 28.41%   R-Sq(adj) = 24.57%

Table 2-22 Comparing four methods of teaching algebra.

Method 1	Method 2	Method 3	Method 4
74	75	76	83
67	69	66	87
81	75	76	84
75	65	79	81
71	78	72	99
69	74	79	71
74	82	74	82
75	78	72	78
70	72	74	78
82	74	79	77
69	77	74	74
71	70	80	68
67	81	71	82
65	68	72	89
63	69	76	78

```
                                 Individual 95% CIs For Mean Based on
                                 Pooled StDev
Level   N    Mean    StDev        - - -+- - - -+- - - -+- - - - + - -
1      15   71.533   5.397       (- - * - -)
2      15   73.800   4.945           (- - * - -)
3      15   74.667   3.792              (- - * - -)
4      15   80.733   7.554                         (- - * - -)
                                 - - +- - - -+- - - -+- - - - + - -
                                 72.0     76.0    80.0       84.0

Pooled StDev = 5.591
```

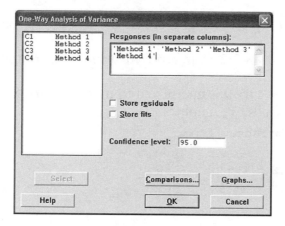

Fig. 2-34.

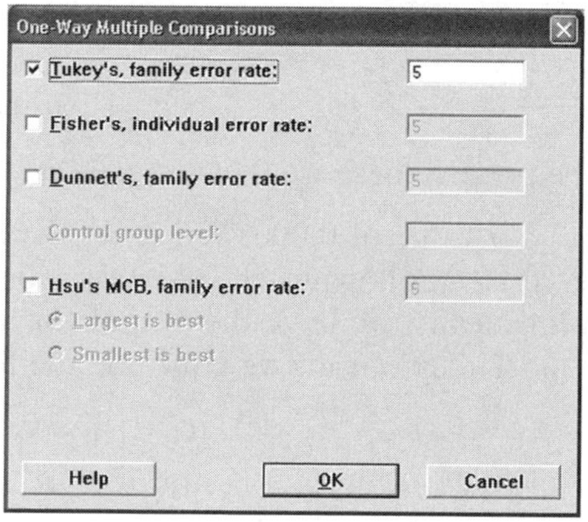

Fig. 2-35.

Tukey 95% Simultaneous Confidence Intervals
All Pairwise Comparisons among Levels of Method

Individual confidence level = 98.94%

Method = 1 subtracted from:

```
Method  Lower    Center   Upper     - - - + - - - -+- - - + - - - + - -
2       -3.132   2.267    7.666          (- -*- -)
3       -2.266   3.133    8.532           (- -*- -)
4        3.801   9.200    14.599                  (- -*- -)
                                     - - - + - - - -+- - - + - - - - + -
                                        7.0      0.0     7.0     14.0
```

$-3.132 < \mu_2 - \mu_1 < 7.666$ is interpreted as $\mu_2 - \mu_1 = 0$, since 0 is in the interval.

$-2.266 < \mu_3 - \mu_1 < 8.532$ is interpreted as $\mu_3 - \mu_1 = 0$, since 0 is in the interval.

$3.801 < \mu_4 - \mu_1 < 14.599$ is interpreted to mean $\mu_4 - \mu_1 > 0$ since the interval contains only positive numbers.

```
Method = 2 subtracted from:

Method  Lower   Center  Upper    - - -+- - - -+- - -+- - - + - -
3       -4.532  0.867   6.266                (- -*- -)
4        1.534  6.933   12.332                  (- -*- -)
                                  - - -+- - - -+- - -+- - - -+ - -
                                  -7.0      0.0     7.0     14.0
```

Following the same logic, $\mu_3 - \mu_2 = 0$ and $\mu_4 - \mu_2 > 0$.

```
Method = 3 subtracted from:
Method  Lower   Center  Upper    - - -+- - - -+- - - -+- - -+ - -
4        0.668  6.067   11.466                  (- -*- -)
                                  - - -+- - - -+- - -+- - - -+ - -
                                  -7.0      0.0     7.0     14.0
```

$\mu_4 - \mu_3$ is taken to be positive because $0.668 < \mu_4 - \mu_3 < 11.466$.

The ANOVA gives a p-value of 0.000. We reject the null hypothesis that the four means are equal. Following the ANOVA output and the 95% confidence intervals for means, we have the output for the Tukey's pairwise comparisons. From the Tukey output we conclude the following:

$$\mu_2 - \mu_1 = 0, \ \mu_3 - \mu_1 = 0, \ \mu_4 - \mu_1 > 0,$$
$$\mu_3 - \mu_2 = 0, \ \mu_4 - \mu_2 > 0, \text{ and } \mu_4 - \mu_3 > 0$$

This is summarized as follows. (Means with common underlining are not different; means without common underlining are different.)

```
Method   1       2       3       4
         _____
```

Suppose five treatments or methods are compared at $\alpha = 0.05$ and the multiple comparison procedure is summarized as follows.

```
Treatment   3       5       2       4       1
            _____
                    _____
```

There are 10 pairs that are compared. The results are as follows. Treatment mean 3 is less than treatment mean 4, treatment mean 3 is less than treatment

mean 1, treatment mean 5 is less than treatment mean 1, treatment mean 2 is less than treatment mean 1. There are no other pairs that are significantly different.

2-6 Exercises for Chapter 2

1. Three different versions of a state tax form as well as the current version are to be compared with respect to the time required to fill out the forms. Forty individuals are selected and paid to participate in the experiment. Ten are randomly assigned to each of four groups. One group fills out form 1, one group fills out form 2, one group fills out form 3, and one group, the current form. The time required by each person in each group is recorded. The data are shown in Table 2-23. The recorded data is the time in hours required to complete the form.

Table 2-23 Time required to fill out four different tax forms.

Form 1	Form 2	Form 3	Current form
3.2	4.4	5.1	4.8
3.9	4.6	4.6	5.6
4.4	3.9	4.1	5.9
4.5	3.2	5.5	5.5
4.0	3.1	4.6	4.8
4.2	4.2	4.4	5.9
4.9	4.7	5.5	5.8
3.9	3.7	4.3	5.1
4.3	3.2	4.4	5.1
4.3	4.5	4.4	6.1

Give the ANOVA for testing that there are no differences in the mean time required to complete the four forms. Test at $\alpha = 0.05$ that there is no difference between the four means. Give the dot plot and box plot comparisons of the four means.

2. Refer to Exercise 1 of this chapter. Perform a Tukey multiple comparison procedure at $\alpha = 0.05$. Summarize your findings by using the underlining technique.

3. Suppose in exercise 1 of this chapter that a block design was used. Four individuals with income less than \$40,000 formed block 1. Similarly four individuals with incomes between \$40,000 and \$50,000 formed block 2, four individuals with incomes between \$50,000 and \$75,000 formed block 3, and four individuals with incomes in excess of \$75,000 formed block 4. The individuals within each block were randomly selected to fill out one of the four forms. The times required to fill out the forms are shown in Table 2-24.

Table 2-24   Time required to fill out four different tax forms for four income groups.

	Form 1	Form 2	Form 3	Current
Group 1	5.5	5.7	6.2	6.5
Group 2	5.0	5.3	5.6	6.0
Group 3	4.5	4.7	5.0	5.5
Group 4	4.0	4.3	5.0	5.3

Give the ANOVA output for a block design. Is there a difference in the time required to fill out the forms for the four blocks? Is there a significant difference in the time required to fill out the four different forms? Which form would you recommend that the state choose?

4. A study was designed to determine the effects of television and Internet connection on student achievement. High school seniors were classified into one of four groups: (1) small time spent watching TV and small time spent on the Internet; (2) small time spent watching TV and large time spent on the Internet; (3) large time spent watching TV and small time spent on the Internet; and (4) large time spent watching TV and large time spent on the Internet. Their cumulative GPA was the response recorded. The results of the study are shown in Table 2-25.

Table 2-25

Internet connect time	Time spent watching TV	
	Small	**Large**
Small	3.9, 4.0, 3.5	2.7, 2.5, 2.8
Large	3.5, 3.3, 3.0	2.0, 2.4, 2.3

Build the ANOVA for this 2 by 2 factorial design. Construct the main effects and interaction graphs. Interpret the results of the experiment.

5. A study was undertaken to determine what combination of products maximized the score that a pizza received. Factor A was cheese and the levels were small and large, factor B was meat and the levels were small and large, and factor C was crust and the levels were thin and thick. Sixteen groups of five people each were randomly assigned to one of the eight combinations and two replications and were asked to assign a score from 0 to 10 (the higher the better) after having eaten the pizza. The response is the average of the five scores of the people comprising the group. The data are given in Table 2-26 (0 is low and 1 is high).

Table 2-26   Effect of factors cheese, meat, and crust on score received by pizzas.

Cheese	Meat	Crust	Rep 1	Rep 2
0	0	0	5.5	6.0
0	0	1	6.0	6.5
0	1	0	8.5	8.7
0	1	1	8.8	9.0
1	0	0	6.2	6.4
1	0	1	6.7	6.6
1	1	0	8.6	8.8
1	1	1	8.3	9.7

Give the ANOVA table for the experiment. Give the interaction and main effect graphs. What is your general recommendation?

6. Fill in the missing blanks within the following ANOVA table (Table 2-27).

Table 2-27

Source of variation	Degrees of freedom	Sum of squares	Mean squares	F-statistic	p-value
Treatments	—	1156	—	—	—
Error	17	—	—		
Total	19	5460			

7. Fill in the missing blanks within the following ANOVA table (Table 2-28).

Table 2-28

Source of variation	Degrees of freedom	Sum of squares	Mean squares	F-statistic	p-value
Treatments	5	—	150	—	—
Blocks	—	500	—	—	—
Error	5	—	50		
Total	15	1500			

8. A 2^2 factorial has been replicated 5 times in a completely randomized design. Fill in the missing blanks within the following ANOVA table (Table 2-29).

9. A 2^3 factorial has been replicated 3 times in a completely randomized design. Fill in the missing blanks within the following ANOVA table (Table 2-30).

Table 2-29

Source of variation	Degrees of freedom	Sum of squares	Mean squares	F-statistic	p-value
A	—	50	—	—	—
B	—	25	—	—	—
AB	—	—	—	—	—
Error	—	350	—		
Total	—	500			

Table 2-30

Source of variation	Degrees of freedom	Sum of squares	Mean squares	F-statistic	p-value
A	—	50	—	—	—
B	—	150	—	—	—
C	—	300	—	—	—
AB	—	15	—	—	—
AC	—	25	—	—	—
BC	—	20	—	—	—
ABC	—	—	—	—	—
Error	—	320	—		
Total	—	885			

10. Tukey's comparison of six treatment means was summarized as follows. Compare all 15 pairs of means a pair at a time.

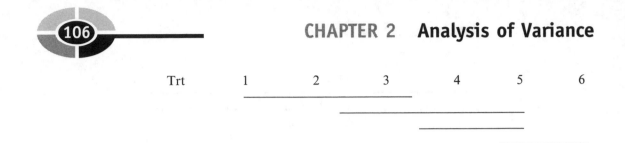

Trt	1	2	3	4	5	6

2-7 Chapter 2 Summary

THE COMPLETELY RANDOMIZED DESIGN

This involves comparing k means when there are k independent samples from normal populations having equal variances. There are n_1 elements from population 1, n_2 elements from population 2,..., n_k elements from population k, and $n = n_1 + n_2 + \cdots + n_k$. The ANOVA is as shown in Table 2-31.

Table 2-31

Source of variation	Degrees of freedom	Sum of squares	Mean squares	F-statistic
Treatments	$k-1$	SST	$\text{MST} = \text{SST}/(k-1)$	$F = \text{MST}/\text{MSE}$
Error	$n-k$	SSE	$\text{MSE} = \text{SSE}/(n-k)$	
Total	$n-1$	SS (total)		

$$\text{SS(total)} = \text{SST} + \text{SSE}$$

Minitab Pull-downs

Stat $\Rightarrow$ ANOVA $\Rightarrow$ Oneway (unstacked) or **Stat $\Rightarrow$ ANOVA $\Rightarrow$ Oneway**

Excel Pull-down

Tools $\Rightarrow$ Data Analysis followed by Anova: Single Factor

THE RANDOMIZED COMPLETE BLOCK DESIGN

There are k treatments randomly assigned within each of b blocks. The ANOVA breakdown for a block design is as shown in Table 2-32.

Table 2-32

Source of variation	Degrees of freedom	Sum of squares	Mean squares	F-statistic
Treatments	$k-1$	SST	$MST = SST/(k-1)$	$F = MST/MSE$
Blocks	$b-1$	SSB	$MSB = SSB/(b-1)$	$F = MSB/MSE$
Error	$n-k-b+1$	SSE	$MSE = SSE/(n-k-b+1)$	
Total	$n-1$	SS(total)		

$$SS(total) = SST + SSB + SSE$$

Minitab Pull-downs

Stat $\Rightarrow$ ANOVA $\Rightarrow$ Two Way

Excel Pull-down

Tools $\Rightarrow$ Data Analysis followed by ANOVA: Two-Factor Without Replication

AN *a* BY *b* FACTORIAL DESIGN

The ANOVA for an *a* by *b* factorial design is as shown in Table 2-33.

Minitab Pull-down

Stat $\Rightarrow$ ANOVA $\Rightarrow$ Balanced ANOVA
Stat $\Rightarrow$ ANOVA $\Rightarrow$ Interactions Plot
Stat $\Rightarrow$ ANOVA $\Rightarrow$ Main effects Plot

Excel Pull-down

Tools $\Rightarrow$ Data Analysis followed by ANOVA: Two-Factor With Replication

Table 2-33

Source of variation	Degrees of freedom	Sum of squares	Mean squares	F-statistic
Factor A	$a-1$	SSA	$MSA = SSA/a-1$	$F = MSA/MSE$
Factor B	$b-1$	SSB	$MSB = SSB/b-1$	$F = MSB/MSE$
Interaction	$(a-1)(b-1)$	SSAB	$MSAB = SSAB/(a-1)(b-1)$	$F = MSAB/MSE$
Error	$n-ab$	SSE	$MSE = SSE/n-ab$	
Total	$n-1$	SS(total)		

$$SS(total) = SSA + SSB + SSAB + SSE$$

A 2^3 FACTORIAL WITH 2 REPLICATIONS

The ANOVA is as shown in Table 2-34.

Table 2-34

Source of variation	Degrees of freedom	Sum of squares	Mean squares	F-statistic
A	1	SSA	MSA	$F = MSA/MSE$
B	1	SSB	MSB	$F = MSB/MSE$
C	1	SSC	MSC	$F = MSC/MSE$
AB	1	SSAB	MSAB	$F = MSAB/MSE$
AC	1	SSAC	MSAC	$F = MSAC/MSE$
BC	1	SSBC	MSBC	$F = MSBC/MSE$
ABC	1	SSABC	MSABC	$F = MSABC/MSE$
Error	8	SSE	MSE	
Total	15			

Simple Linear Regression and Correlation

3-1 Probabilistic Models

Deterministic models describe the connection between independent variables and a dependent variable. They are so named because they allow us to determine the value of the dependent value from the values of the independent variables. These deterministic models are usually from the natural sciences. Some examples of deterministic models are as follows:

- $E = mc^2$, where $E =$ energy, $m =$ mass, and $c =$ the speed of light
- $F = ma$, where $F =$ force, $m =$ mass, and $a =$ acceleration

- $S = at^2/2$, where S = distance, t = time, and a = gravitational acceleration
- $D = vt$, where D = distance, v = velocity, and t = time.

An automobile traveling at a constant speed of fifty miles per hour will travel the distances shown in Table 3-1 for the given times.

Probabilistic models are more realistic for most real-world situations. For example, suppose we know that, in a given city, most lots sell for about $20,000 and that the cost of building a new house costs about $70 per square foot. The average cost, y, of a house of 2500 square feet is

$$y = 20,000 + 70(2500) = \$195,000$$

This is still a deterministic model. We know for example that the cost of twenty homes, each of 2500 square feet, would likely vary. The actual costs might be given by the twenty costs in Table 3-2.

A more reasonable model would be

$$y = 20,000 + 70(2500) + \varepsilon$$

ε is a random variable, and is called a *random error*.

Table 3-1

D	t
50 miles	1 hour
100 miles	2 hours
150 miles	3 hours
200 miles	4 hours
250 miles	5 hours

Table 3-2

192 000	182 000	17 0000	178 000	183 000
199 000	202 000	206 000	203 000	205 000
194 000	195 000	202 250	195 000	202 500
187 000	174 000	202 250	206 000	199 000

Table 3-3

−3000	−13 000	−25 000	−17 000	−12 000
4000	7000	11 000	8000	10 000
−1000	0	7250	0	7500
−8000	−21 000	7250	11 000	4000

The values of ε for the twenty homes in Table 3-2 are given in Table 3-3.

Generalizing, the cost for a house of x square feet in size is given by the probabilistic model

$$y = 20{,}000 + 70x + \varepsilon$$

The general form of a probabilistic model is

$$y = \text{deterministic component} + \text{random error component}$$

We always assume that the mean value of the random error equals 0. This is equivalent to assuming that the mean value of y, $E(y)$, equals the deterministic component.

$$E(y) = \text{deterministic component}$$

In this chapter, the deterministic component will be a straight line, written as $\beta_0 + \beta_1 x$. Fitting this model to a data set is an example of *regression modeling* or *regression analysis*.

Summarizing, a *first-order (straight-line) probabilistic model* is given by

$$y = \beta_0 + \beta_1 x + \varepsilon$$

y is the dependent or response variable, and x is the independent or predictor variable. The expression $E(y) = \beta_0 + \beta_1 x$ is referred to as the *line of means*.

As an example, suppose there is a population of rodents and that the adult male weights, y, are related to the heights, x, by the relationship $y = 0.4x - 5.2 + \varepsilon$. The height x is in centimeters and the weight y is in kilograms. The error component is normally distributed with mean equal to 0 and standard deviation 0.1. Note that the population relationship is usually not known, but we are assuming it is known here to develop the concepts. In fact, we are usually trying to establish the relationship between y and x. We capture ten of these rodents and determine their heights and weights. This data is given in Table 3-4, and a plot is shown in Fig. 3-1.

Table 3-4 The heights and weights of 10 rodents.

Rodent #	Height, x	Weight, y
1	14.5	0.69
2	14.5	0.52
3	15.0	0.93
4	15.0	0.65
5	15.0	0.97
6	15.4	0.95
7	15.4	1.05
8	15.4	0.95
9	15.5	1.03
10	15.5	0.99

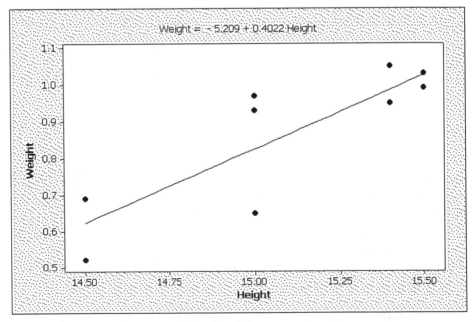

Fig. 3-1.

All rodents that are 14.5 cm tall have an average weight of $0.4(14.5) - 5.2 = 0.6$ kg, all rodents that are 15.0 cm tall have an average weight of $0.4(15.0) - 5.2 = 0.8$ kg, and so forth. The actual captured rodents have weights that vary about the line of means. Also note that the taller the rodent, the heavier it is.

The concepts of regression are sophisticated and an effort has been made to help the reader understand these concepts. As mentioned earlier, we do not usually know the equation of the deterministic line that connects y with x. What we shall see in the next section is that we can sample the population and gather a set of data such as that shown in the height–weight table above and estimate the deterministic equation.

The *assumptions of regression* are: (1) Normality of error. The error terms are assumed to be normally distributed with a mean of zero for each value of x. (2) The variation around the line of regression is constant for all values of x. This means that the errors vary by the same amount for small x as for large x. (3) The errors are independent for all values of x.

3-2 The Method of Least Squares

Purpose: The purpose of the least-squares method is to find the equation of the straight line that fits the data best in the sense of least squares. Calculus techniques are used to find the equation.

Assumptions: The assumptions of regression.

The relationship between the number of hours studied, x, and the score, y, made on a mathematics test is postulated to be linear. The linear model $y = \beta_0 + \beta_1 x + \varepsilon$ is proposed. Ten students are sampled and the scores and hours studied are recorded as in Table 3-5.

A scatter plot (Fig. 3-2) is drawn using Minitab. The pull-down is **Graph ⇒ Scatterplot**.

The scatter plot shows a clear linear trend. The population equation is $y = \beta_0 + \beta_1 x + \varepsilon$. Since we have only a sample of all possible values for x and y, we can at best estimate the deterministic part of the model $E(y) = \beta_0 + \beta_1 x$. The notation for the estimate is $\hat{y} = b_0 + b_1 x$. By using the data in the table and some calculus we can derive an expression for b_0, the estimate of β_0, and for b_1, the estimate of β_1. If we define SS_{xx} and SS_{xy} as follows

$$SS_{xx} = \sum (x - \bar{x})^2 \quad \text{and} \quad SS_{xy} = \sum (x - \bar{x})(y - \bar{y})$$

Table 3-5

Student	Hours studied	Score on test
Long	10	78
Reed	15	83
Farhat	8	75
Konvalina	7	77
Wileman	13	80
Maloney	15	85
Kidd	20	95
Carter	10	83
Maher	5	65
Carroll	5	68

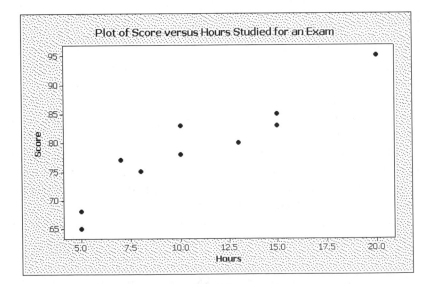

Fig. 3-2.

then the expression for b_1 is

$$b_1 = \frac{SS_{xy}}{SS_{xx}}$$

and the expression for b_0 is

$$b_0 = \bar{y} - b_1\bar{x}$$

The Minitab software or the Excel software will evaluate b_0 and b_1 when the data are supplied. The data for x and y are entered into columns C1 and C2 of the Minitab worksheet. The pull-down **Stat ⇒ Regression ⇒ Fitted Line Plot** gives the data and the fitted line as shown in Fig. 3-3. The equation for the estimated regression line is shown at the top of this figure, where the equation $\hat{y} = b_0 + b_1x$ is given as (approximately) Score $= 61.23 + 1.64$ hours. That is, $b_0 = 61.23$ and $b_1 = 1.64$. The slope $b_1 = 1.64$ tells us that, for each additional hour of study, the score increased by 1.64. The y-intercept $b_0 = 61.23$ is the value of the score when no hours were spent on study. Since 0 is outside the range of hours studied, it does not have an interpretation in the context of the scores.

Table 3-6 compares the observed and predicted values. Except for round-off errors, the sum of column 4 will always be zero and the calculus ensures us that SSE $= 85.591$ will be the smallest it can be for any straight line fitted to the data. That is, if any line other than $\hat{y} = 61.2273 + 1.63636x$ is fit to the data and SSE computed, it will be larger than 85.591. The values

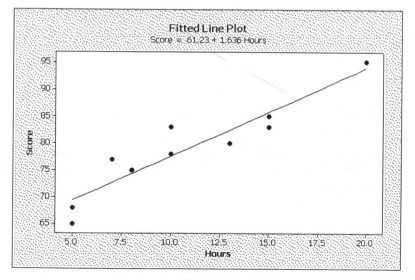

Fig. 3-3.

Table 3-6

x	Observed y	Predicted $\hat{y}=61.2273+1.63636x$	$(y-\hat{y})$	$(y-\hat{y})^2$
10	78	77.5909	0.40910	0.1674
15	83	85.7727	−2.77270	7.6879
8	75	74.3182	0.68182	0.4649
7	77	72.6818	4.31818	18.6467
13	80	82.5000	−2.49998	6.2499
15	85	85.7727	−0.77270	0.5971
20	95	93.9545	1.04550	1.0931
10	83	77.5909	5.40910	29.2584
5	65	69.4091	−4.40910	19.4402
5	68	69.4091	−1.40910	1.9856
			Sum $= 0.00012$	SSE $= 85.591$

$b_1 = SS_{xy}/SS_{xx}$ and $b_0 = \bar{y} - b_1\bar{x}$ for the slope and intercept minimize the sums of squares.

EXAMPLE 3-1

In a recent *USA Today* article entitled "Vidal Sassoon takes on a hairy fight against P&G," Vidal Sassoon said the consumer giant mishandled his life's work, while Procter and Gamble said the product line lost its cachet. Table 3-7 shows the Vidal Sassoon product sales in the USA in millions for the years 1998 till 2003. Assuming the trend continues into 2004, predict the US sales for 2004.

SOLUTION

A plot of the data is shown in Fig. 3-4. Suppose we code the years as 1, 2, 3, 4, 5, and 6. From Minitab, we obtain the estimated regression line **Sales $= 82.7 - 11.0$ year (coded)**. Assuming the trend continues, the

Table 3-7

Year	Year coded	Sales in millions
1998	1	71.3
1999	2	59.5
2000	3	51.9
2001	4	41.1
2002	5	24.9
2003	6	17.5

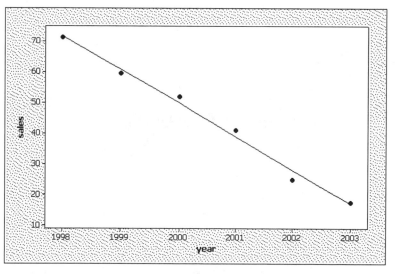

Fig. 3-4.

estimated sales for 2004 amount to $82.7 - 11.0 \ (7) = 5.7$ million. Note that the slope of this line is negative since, as the years increase, the sales are decreasing. It is estimated that as the year increases by one the sales decrease by 11 million.

EXAMPLE 3-2
Give the Excel solution to Example 3-1.

SOLUTION
The Excel solution to this example proceeds as follows. The coded years and sales are entered into columns A and B. The pull-down **Tools** ⇒ **Data Analysis** produces the Data Analysis dialog box and we select Regression, as shown in Fig. 3-5.

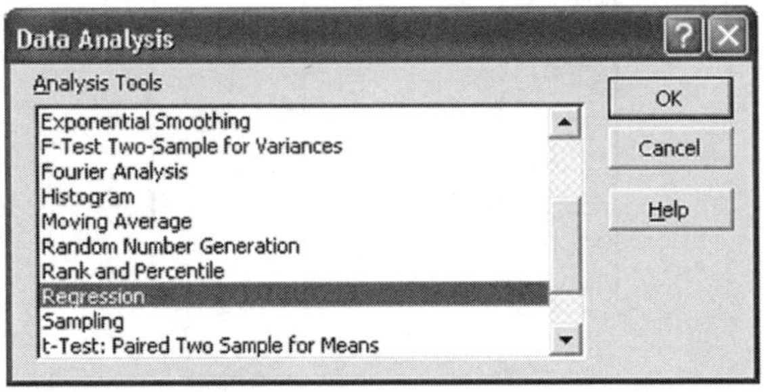

Fig. 3-5.

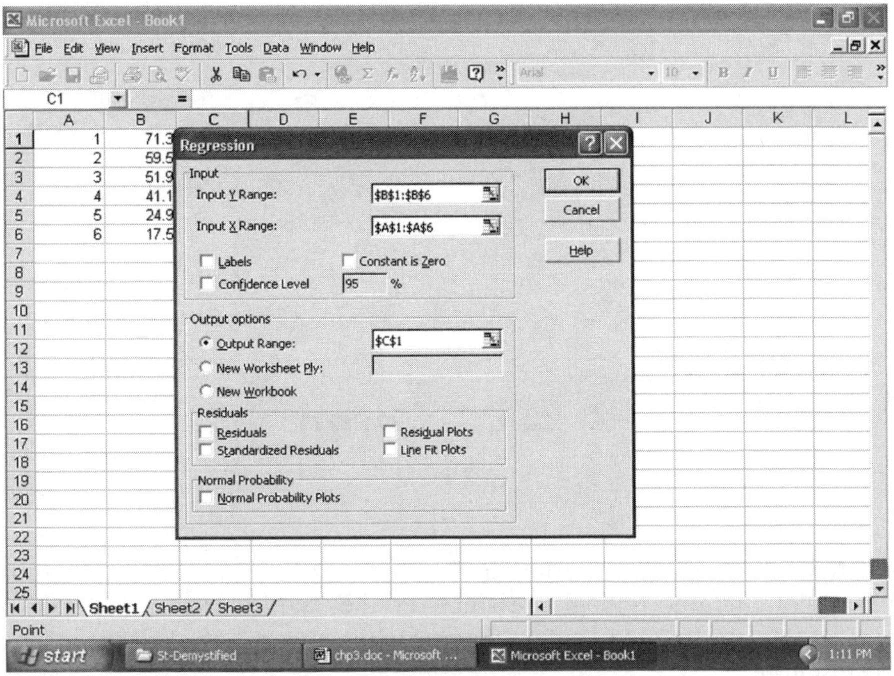

Fig. 3-6.

Figure 3-6 shows the Regression dialog box filled in. Figure 3-7 gives the output for Excel. The value for b_0 is under the Coefficients column as intercept and the slope of the regression line, b_1, is under the Coefficients column as X Variable 1. The equation of the line of best fit is $\hat{y} = 82.7267 - 10.96x$.

	A	B	C	D	E	F	G	H	I	J	K
1	1	71.3	SUMMARY OUTPUT								
2	2	59.5									
3	3	51.9	Regression Statistics								
4	4	41.1	Multiple R	0.995163							
5	5	24.9	R Square	0.990349							
6	6	17.5	Adjusted R Square	0.987936							
7			Standard Error	2.263036							
8			Observations	6							
9											
10			ANOVA								
11				df	SS	MS	F	ignificance F			
12			Regression	1	2102.128	2102.128	410.465	3.5E-05			
13			Residual	4	20.48533	5.121333					
14			Total	5	2122.613						
15											
16				Coefficient	andard Err	t Stat	P-value	Lower 95%	Upper 95%		
17			Intercept	82.72667	2.106772	39.26702	2.51E-06	76.87732	88.57602		
18			X Variable 1	-10.96	0.540969	-20.2599	3.5E-05	-12.462	-9.45803		

Fig. 3-7.

3-3 Inferences About the Slope of the Regression Line

Purpose of the test: The purpose of the test is to determine whether a given value is reasonable for the slope of the population regression line (H_0: $\beta_1 = c$). The test H_0: $\beta_1 = 0$ is a test to determine whether a straight line should be fit to the data. If the null hypothesis is not rejected, then a straight line does not model the relationship between x and y.

Assumptions: The assumptions of regression.

The regression model is $y = \beta_0 + \beta_1 x + \varepsilon$ and β_1 is the slope of the model. The slope of the model tells you how y changes with a unit change in x. To test the null hypothesis that β_1 equals some value, say c, we divide the difference $(b_1 - c)$ by the standard error of b_1. The following test statistic is used to test H_0: $\beta_1 = c$ versus H_a: $\beta_1 \neq c$

$$t = \frac{b_1 - c}{\text{standard error of } b_1}$$

t has a Student t distribution with $(n - 2)$ degrees of freedom.

EXAMPLE 3-3
Table 3-8 give the systolic blood pressure readings and weights for 10 newly diagnosed patients with high blood pressure. A plot of systolic blood pressure versus weight is shown as a Minitab output in Fig. 3-8.

Table 3-8

Patient	Systolic	Weight
Jones	145	210
Konvalina	155	245
Maloney	160	260
Carrol	155	230
Wileman	130	175
Long	140	185
Kidd	135	230
Smith	165	249
Hamilton	150	200
Bush	130	190

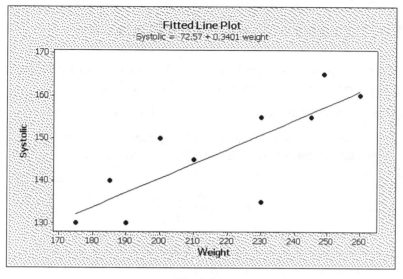

Fig. 3-8.

SOLUTION

```
The regression equation is
Systolic = 72.6 + 0.340 weight
```

Predictor	Coef	SE Coef	T	P
Constant	72.57	19.46	3.73	0.006
weight	**0.34007**	**0.08878**	3.83	0.005

Suppose we wished to test that the systolic blood pressure increases one point for each pound that the patient's weight increases. The test statistic is computed as follows. (Note that $c = 1$, $b_1 = 0.34007$, and the standard error of $b_1 = 0.08878$.)

$$t = \frac{0.34 - 1.0}{0.089} = -7.42$$

At $\alpha = 0.05$, the t values with 8 degrees of freedom are ± 2.306. The data would refute the null hypothesis. Each additional pound would increase the systolic pressure by less than 1. Note: The T value (3.83) shown in the output above along with the two-tailed p-value (0.005) is for the null hypothesis H_0: $\beta_1 = 0$ versus H_a: $\beta_1 \neq 0$.

3-4 The Coefficient of Correlation

Purpose: The purpose of the correlation coefficient is to measure the strength of the linear relationship between two random variables.

Assumptions: The random variables X and Y have a bivariate distribution.

A measure very much related to the slope of the regression line is the Pearson correlation coefficient. The sample Pearson correlation coefficient is defined to be

$$r = \frac{SS_{xy}}{\sqrt{SS_{xx}SS_{yy}}}$$

It is measured on a sample and is an estimate of the population Pearson correlation coefficient ρ. Recall that the estimated slope of the regression line is

$$b_1 = \frac{SS_{xy}}{SS_{xx}}$$

Note that $r = 0$ if and only if $b_1 = 0$. The main difference between the two is that r ranges between -1 and $+1$ and b_1 does not. The correlation coefficient measures the strength of the linear relationship between x and y. If the points fall on a straight line with a positive slope, then $r = 1$. If the points fall on a straight line with a negative slope, then $r = -1$. If the points form a shotgun pattern, the value of $r = 0$. Most points don't fall on a straight line, but indicate a positive or negative trend.

EXAMPLE 3-4
Consider the following systolic blood pressure–weight readings for ten individuals (Table 3-9).

SOLUTION
The values are put into columns C1 and C2 of the Minitab worksheet and the pull-down **Stat** $\Rightarrow$ **Basic Statistics** $\Rightarrow$ **Correlation** is performed. The following Minitab output results.

> **Correlations: Systolic, Weight**
> *Pearson correlation of Systolic and Weight = 0.804*
> *P-Value = 0.005*

The correlation coefficient is 0.804 and the p-value corresponding to the null hypothesis of no correlation (H_0: $\rho = 0$) is 0.005. There is a positive

Table 3-9

Patient	Systolic, y	Weight, x
Jones	145	210
Konvalina	155	245
Maloney	160	260
Carrol	155	230
Wileman	130	175
Long	140	185
Kidd	135	230
Smith	165	249
Hamilton	150	200
Bush	130	190

correlation in the population between weight and systolic blood pressure, at $\alpha = 0.05$. The plot of the data is shown in Fig. 3-8 and shows a linear trend.

EXAMPLE 3-5
Find the correlation coefficient using Excel.

SOLUTION
If Excel is used to compute the correlation coefficient, the pull-down **Tools ⇒ Data Analysis** is used and correlation is selected from the Data Analysis dialog box. The correlation dialog box is filled in as in Fig. 3-9.

The following output is obtained. Note that the correlation coefficient is 0.804464.

C	D	E
	Systolic	Weight
Systolic	1	
Weight	0.804464	1

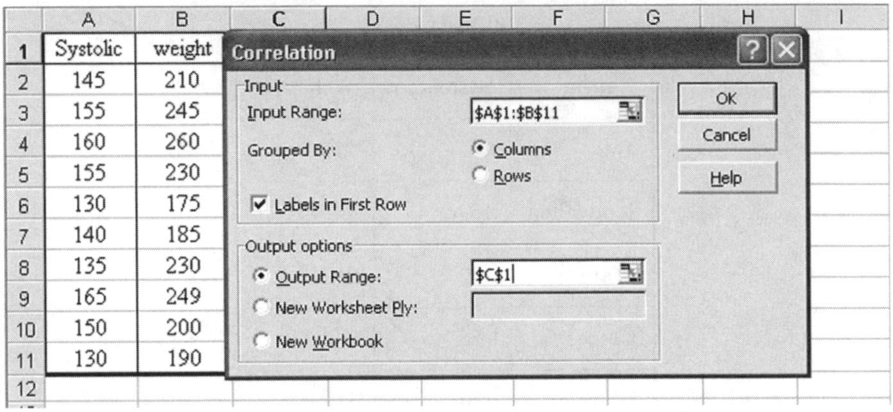

	A	B	C	D	E	F	G	H	I
1	Systolic	weight	Correlation					?✕	
2	145	210	Input					OK	
3	155	245	Input Range:		$A$1:$B$11				
4	160	260	Grouped By:		⦿ Columns			Cancel	
5	155	230			○ Rows			Help	
6	130	175	☑ Labels in First Row						
7	140	185							
8	135	230	Output options						
9	165	249	⦿ Output Range:		$C$1				
10	150	200	○ New Worksheet Ply:						
11	130	190	○ New Workbook						
12									

Fig. 3-9.

The following example illustrates two variables that are negatively correlated.

EXAMPLE 3-6
In Table 3-10, the dependent variable is the cumulative GPA (Grade Point Average) and the independent variable is the number of hours per week of TV watched. The sample consists of fifteen high school freshers.

Table 3-10

TV hours	10	5	15	16	26	23	10	30	13	8	27	22	28	32	12
GPA	3.7	3.5	3.0	3.1	2.8	2.6	2.8	2.2	3.6	3.4	2.5	3.1	2.4	2.0	3.4

SOLUTION
A plot of the data is shown in Fig. 3-10. The correlation coefficient as computed by Minitab and Excel is $r = -0.875$. The plot shows the negative linear relationship between the two variables. The value of r indicates the strength of the relationship.

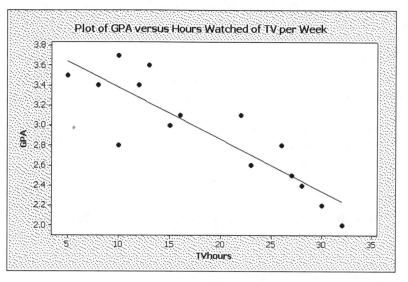

Fig. 3-10.

3-5 The Coefficient of Determination

Purpose: The purpose of this coefficient is to measure the strength of the linear relationship between the dependent variable y, and the predictor variable x.

Assumptions: The assumptions of regression.

The *analysis of variance* (ANOVA) for simple linear regression may be represented as in Table 3-11.

Table 3-11

Source	d.f.	Sums of squares	Mean squares	F-value
Explained variation	1	SSR	$MSR = SSR/1$	$F = MSR/MSE$
Unexplained variation	$n-2$	SSE	$MSE = SSE/(n-2)$	
Total	$n-1$	SS(total)		

The symbol r^2 is used to represent the ratio SSR/SS(total) and is called the coefficient of determination. The coefficient of determination measures the proportion of variation in y that is explained by x.

> Note that the coefficient of determination may also be found by squaring the correlation coefficient. Also, the source *explained variation* is also called *regression variation* and *unexplained variation* is also called *residual variation*.

EXAMPLE 3-7
A study was conducted concerning the contraceptive prevalence (x) and the fertility rate (y) in developing countries. The data is shown in Table 3-12.

SOLUTION
The Minitab output is as follows.

S = 0.7489   **R-Sq = 65.8%**   R-Sq(adj) = 61.5%

Table 3-12

Country	Contraceptive prevalence, x	Fertility rate, y
Thailand	69	2.3
Costa Rica	71	3.5
Turkey	62	3.4
Mexico	55	4.0
Zimbabwe	46	5.4
Jordan	35	5.5
Ghana	14	6.0
Pakistan	13	5.0
Sudan	10	4.8
Nigeria	7	5.7

Analysis of Variance

Source	DF	SS	MS	F	P
Regression	1	**8.6171**	8.6171	15.36	0.004
Residual Error	8	4.4869	0.5609		
Total	9	**13.1040**			

The coefficient of determination is shown as **R-Sq = 65.8%**. Alternatively, it may be computed as

$$r^2 = \frac{SSR}{SS(total)} = \frac{8.6167}{13.1040} \times 100 = 65.8\%$$

EXAMPLE 3-8
Solve Example 3-7 using Excel.

SOLUTION
The Excel output is shown in Fig. 3-11. The coefficient of determination from the Excel worksheet is shown as **R Square 0.657591**. The interpretation

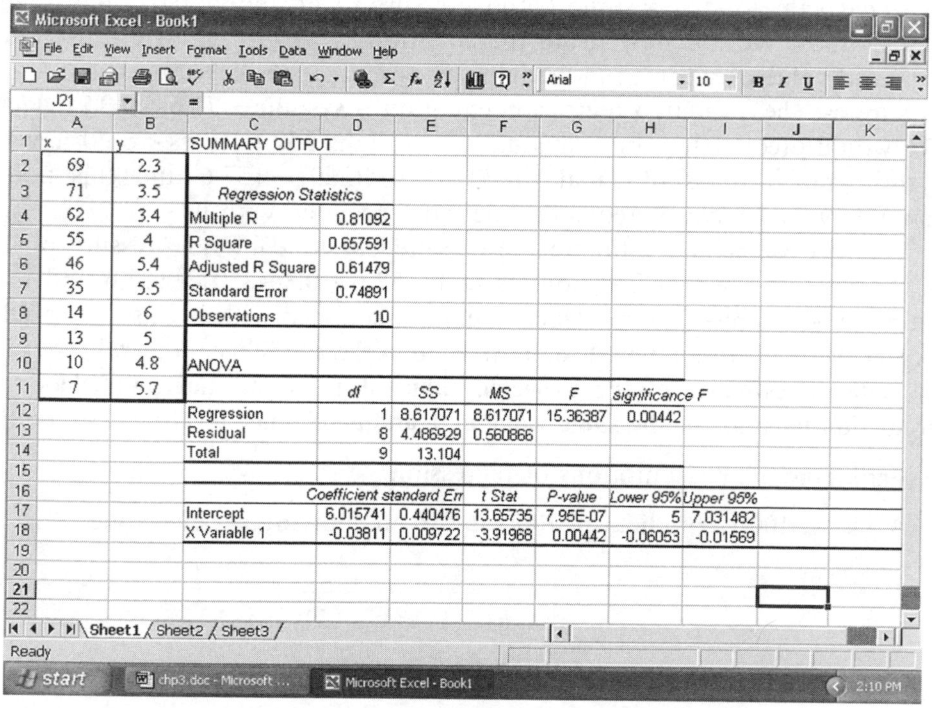

Fig. 3-11.

is that about 66% of the variation in fertility rates is explainable by the variation in contraceptive prevalence.

3-6 Using the Model for Estimation and Prediction

Purpose: The estimated regression equation $\hat{y} = b_0 + b_1 x$ can be used to predict the value of y for some value of x, say x_0, or the same equation can be used to estimate the mean value of the ys corresponding to x_0. For example, consider a regression study where y represents systolic blood pressure and x represents weight. If we use the estimated regression equation to predict the systolic blood pressure of an individual who weighs $x_0 = 250$ pounds we are using the equation $\hat{y} = b_0 + b_1 x$ to predict the particular value of y for a given x, say $x_0 = 250$.

Now suppose we wish to use the regression equation to estimate the expected value of the systolic blood pressure of all individuals who weigh 250 pounds. We are now using the equation $\hat{y} = b_0 + b_1 x$ to estimate the expected value of y for all individuals who weigh $x_0 = 250$ pounds.

Suppose the estimated regression equation is **systolic $= 72.6 + 0.34$ weight**. We would predict that an individual who weighs 250 pounds would have a systolic blood pressure equal to $72.6 + 0.34(250) = 157.6$ and this point estimate would have a prediction interval associated with it.

Likewise we would estimate the expected systolic blood pressure of all individuals who weigh 250 pounds to be $72.6 + 0.34(250) = 157.6$ and this estimate would have a confidence interval associated with it. **Also, we would expect the prediction interval to be wider than the confidence interval. That is, the interval estimate of the expected value of y will be narrower than the prediction interval for the same value of x and confidence level.**

Assumptions: The assumptions of regression.

A $(1 - \alpha)100\%$ prediction interval for an individual new value of y at $x = x_0$ is

$$\hat{y} \pm t_{\alpha/2}\, s \sqrt{1 + \frac{1}{n} + \frac{(x_0 - \bar{x})^2}{SS_{xx}}}$$

where $\hat{y} = b_0 + b_1 x_0$, the t value is based on $(n - 2)$ degrees of freedom, $s = \sqrt{SSE/(n - 2)}$ and is referred to as the estimated standard error of the regression model, n is the sample size, x_0 is the fixed value of x, $\bar{x}$ is the mean of the observed x values, and $SS_{xx} = \sum (x - \bar{x})^2$.

A $(1 - \alpha)100\%$ confidence interval for the mean value of y at $x = x_0$ is

$$\hat{y} \pm t_{\alpha/2}s\sqrt{\frac{1}{n} + \frac{(x_0 - \bar{x})^2}{SS_{xx}}}$$

and the parts of the equation are as defined as in the prediction interval discussion. Note that the only difference in the two expressions is the additional 1 under the square root in the prediction interval. This additional 1 makes the prediction interval wider.

EXAMPLE 3-9

A study was conducted on 15 diabetic patients. The independent variable x was the hemoglobin A1C value, taken after three months of taking the fasting blood glucose value each morning of the three-month period and averaging the values. The latter value was the dependent variable value y. The data were as shown in Table 3-13.

We wish to set a 95% prediction interval for the average glucose reading of a diabetic who has a hemoglobin A1c value of 7.0 as well as a 95% confidence interval for all diabetics with a hemoglobin A1c value of 7.0. The data are entered in the Minitab worksheet as shown in Fig. 3-12.

SOLUTION

The pull-down **Stat** $\Rightarrow$ **Regression** $\Rightarrow$ **Regression** gives the dialog box shown in Fig. 3-13. Choosing the options box in Fig. 3-13 and filling it in as shown in Fig. 3-14, the following Minitab output is obtained.

```
The regression equation is
y = 60.6 + 10.4 x

Predictor      Coef           SE Coef          T            P
Constant       60.552         7.476            8.10         0.000
x              10.4056        0.9743           10.68        0.000

S = 5.881             R-Sq = 89.8%        R-Sq(adj) = 89.0%


Analysis of Variance

Source            DF     SS        MS       F        P
Regression         1     3945.3    3945.3   114.08   0.000
Residual Error    13      449.6      34.6
Total             14     4394.9
```

Table 3-13

Patient	x, hemoglobin A1c	y, average fasting blood sugar over 3-month period
Jones	6.1	120
Liu	6.8	146
Maloney	6.5	125
Reed	7.1	135
Smith	7.4	140
Lee	5.8	115
Aster	8.0	145
Bush	8.3	147
Carter	8.0	150
Haley	5.5	110
Long	10.0	160
Carroll	7.7	145
Maher	9.0	155
Grobe	11.0	170
Lewis	5.5	118

```
Unusual Observations

Obs     x        y        Fit      SE Fit   Residual   St Resid
 2      6.8    146.00   131.31     1.67     14.69       2.61R
14     11.0    170.00   175.01     3.72     -5.01      -1.10 X
```

R denotes an observation with a large standardized residual
X denotes an observation whose X value gives it large influence.

↓	C1	C2	C3
	x	y	
1	6.1	120	
2	6.8	146	
3	6.5	125	
4	7.1	135	
5	7.4	140	
6	5.8	115	
7	8.0	145	
8	8.3	147	
9	8.0	150	
10	5.5	110	
11	10.0	160	
12	7.7	145	
13	9.0	155	
14	11.0	170	
15	5.5	118	

Fig. 3-12.

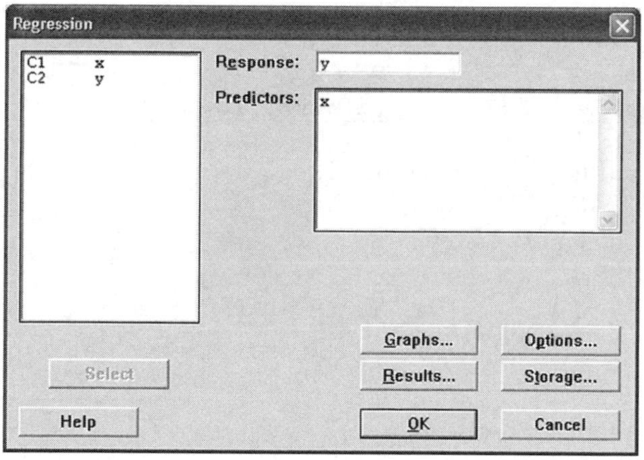

Fig. 3-13.

Predicted Values for New Observations

New Obs	Fit	SE Fit	95.0% CI	95.0% PI
1	133.39	1.60	(129.94, 136.85)	(120.23, 146.56)

Values of Predictors for New Observations

New Obs	x
1	7.00

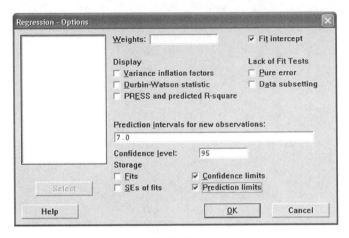

Fig. 3-14.

The part of the output that contains the prediction and the confidence interval is

Predicted Values for New Observations

New Obs	Fit	SE Fit	95.0% CI	95.0% PI
1	133.39	1.60	(129.94, 136.85)	(120.23, 146.56)

The Fit value is $\hat{y} = 60.6 + 10.4(7.0) = 133.39$. The SE Fit is $s\sqrt{1/n + (x_0 - \bar{x})^2/SS_{xx}}$ or

$$5.881\sqrt{\frac{1}{15} + \frac{(7.0 - 7.531)^2}{36.425}} = 1.604$$

The 95% confidence interval is $\hat{y} \pm t_{\alpha/2}\, s\sqrt{(1/n) + (x_0 - \bar{x})^2/SS_{xx}}$ or

$$133.39 \pm 2.160(5.881)\sqrt{\frac{1}{15} + \frac{(7.0 - 7.531)^2}{36.425}}$$

or $133.39 \pm 2.160(1.604)$ or 133.39 ± 3.46 or $(129.93, 136.85)$. Similarly the 95% prediction interval can be shown to be $(120.23, 146.56)$.

Summarizing, we are 95% confident that a diabetic with a hemoglobin A1c value of 7.0 had a fasting blood sugar over the past three months that averaged between 120.23 and 146.56. We are 95% confident that diabetics with a hemoglobin A1c value of 7.0 had an average fasting blood sugar over the past three months between 129.93 and 136.85.

EXAMPLE 3-10
Solve Example 3-9 using Excel.

SOLUTION
If Excel is used to solve the same problem, we would proceed as follows. Enter the data into the Excel worksheet. Use the pull-down **Tools** $\Rightarrow$ **Data Analysis** to access the Data Analysis dialog box. Select Regression in order to perform a regression analysis. Fill out the Regression dialog box as shown in Fig. 3-15. This will produce the output given in Fig. 3-16.

Now suppose we wish to form a 95% prediction interval and a 95% confidence interval for y when $x = 7.0$. The two intervals are

$$\hat{y} \pm t_{\alpha/2}\, s\sqrt{1 + \frac{1}{n} + \frac{(x_0 - \bar{x})^2}{SS_{xx}}} \quad \text{and} \quad \hat{y} \pm t_{\alpha/2}\, s\sqrt{\frac{1}{n} + \frac{(x_0 - \bar{x})^2}{SS_{xx}}}$$

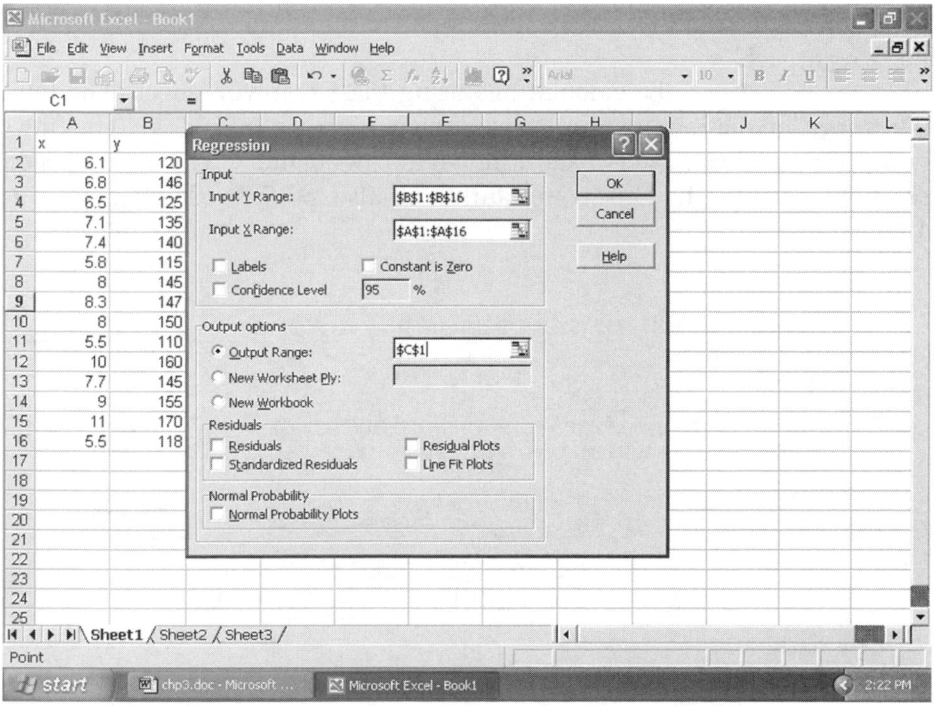

Fig. 3-15.

	A	B	C	D	E	F	G	H	I
1	x	y	SUMMARY OUTPUT						
2	6.1	120							
3	6.8	146	*Regression Statistics*						
4	6.5	125	Multiple R	0.94747					
5	7.1	135	R Square	0.897699					
6	7.4	140	Adjusted R Square	0.88983					
7	5.8	115	Standard Error	5.8809					
8	8	145	Observations	15					
9	8.3	147							
10	8	150	ANOVA						
11	5.5	110		df	SS	MS	F	Significance F	
12	10	160	Regression	1	3945.329	3945.329	114.0763	8.33726E-08	
13	7.7	145	Residual	13	449.6048	34.58499			
14	9	155	Total	14	4394.933				
15	11	170							
16	5.5	118		Coefficients	Standard Err	t Stat	P-value	Lower 95%	Upper 95%
17			Intercept	60.55238	7.475701	8.099893	1.95E-06	44.40211408	76.70265
18			x	10.40563	0.97425	10.68065	8.34E-08	8.300888707	12.51037
19									

Fig. 3-16.

$\hat{y} = 60.6 + 10.4(7.0) = 133.39$ is the point estimate, $n = 15$, $x_0 = 7.0$, $\bar{x} = 7.531$, $s = 5.881$ from Fig. 3-16 and is called standard error, $t_{.025} = 2.160$, and $SS_{xx} = 36.425$.

Figure 3-17 shows the calculations using the Excel worksheet. The lower and upper prediction intervals are shown in rows 3 and 4. The lower and upper confidence intervals are shown in rows 6 and 7. The values needed are shown in row 1. This computation is also performed when you use Minitab.

H7		=	=133.39+2.16*5.881*SQRT((1/15)+(7-7.531)^2/36.425)						
	A	B	C	D	E	F	G	H	I
1	yhat = 133.39	n = 15	x0 = 7.0	xbar = 7.531	s=5.881	t=2.160	SS=36.425		
2									
3	lower prediction = 133.39 - 2.160*5.881*sqrt(1 + (1/15)+(7-7.531)^2/36.425)							120.2229	
4	upper prediction= 133.39+ 2.160*5.881*sqrt(1 + (1/15)+(7-7.531)^2/36.425)							146.5571	
5									
6	lower confidence=133.39-2.160*5.881*sqrt((1/15)+(7-7.531)^2/36.425)							129.9249	
7	upper confidence=133.39+2.160*5.881*sqrt((1/15)+(7-7.531)^2/36.425)							136.8551	
8									

Fig. 3-17.

3-7 Exercises for Chapter 3

1. Give the deterministic equation for the line passing through the following pairs of points: (a) $(1, 1.5)$ and $(3, 8.5)$; (b) $(0, 1)$ and $(2, -3)$; (c) $(0, 3.1)$ and $(1, 4.8)$.

2. Give the slope and y-intercept of the deterministic equations in problem 1.

3. Find the equation of the line that fits the following set of data (Table 3-14) best in the least-squares sense.

Table 3-14

x	y
1.3	20.3
2.7	22.5
3.2	25.4
4.1	26.3
6.2	33.6

4. Find the equation of the line that fits the following set of data (Table 3-15) best in the least-squares sense.

Table 3-15

x	y
1.0	3.7
2.0	3.3
3.0	2.8
4.0	3.1
5.0	1.7

5. A study was conducted to find the relationship between alcohol consumption and blood pressure. The systolic blood pressure, y, and the number of drinks per week, x, were recorded for a group of ten patients with poorly controlled high blood pressure. The data were as shown in Table 3-16.

Table 3-16

y	x
145	7
155	10
140	6
166	21
156	12
160	15
145	10
170	18
135	14
150	17

(a) Does a plot suggest a linear relationship between x and y?
(b) If so, is it positive or negative?
(c) Find values for b_0 and b_1 and interpret their values.

6. In problem 5, test whether there is a positive relationship between alcohol consumption and blood pressure. That is, test H_0: $\beta_1 = 0$ versus H_a: $\beta_1 > 0$ at $\alpha = 0.05$.

7. Calculate the correlation coefficient between alcohol consumption and blood pressure for the data in problem 5. Also, give the coefficient of determination and interpret it.

8. Give a 95% prediction interval and a 95% confidence interval for poorly controlled high blood pressure patients who consume 20 alcoholic drinks per week.

9. A cell phone company looked at the ages (x) and the number of cell phone calls placed per month by twenty of its subscribers (y). The data are as shown in Table 3-17.

Table 3-17

x	y	x	y
18	75	35	35
26	50	40	30
30	45	17	125
45	35	18	124
55	15	25	49
19	107	27	79
27	46	33	40
20	79	43	30
24	59	23	60
25	55	58	24

(a) Find the regression line connecting the number of calls to the age of the subscriber.

(b) Find the 95% prediction interval and the 95% confidence interval for the number of calls placed by a 20 year old.

(c) Give the adjusted coefficient of determination.

(d) Give the correlation coefficient between the number of calls and the age of the caller.

10. A psychologist obtains the flexibility and creativity scores on nine randomly selected mentally retarded children. The results are given in Table 3-18.

Table 3-18

Flexibility x	Creativity y
3	2
4	6
3	4
6	7
9	13
6	8
5	5
5	8
7	10

(a) Find the regression line connecting the creativity score to the flexibility score.
(b) Find the 95% prediction interval and the 95% confidence interval for the creativity score of a mentally retarded child who scored 8 on flexibility.
(c) Give the adjusted coefficient of determination.
(d) Give the correlation coefficient between the flexibility score and the creativity score.

3-8 Chapter 3 Summary

The *estimated regression line* is

$$\hat{y} = b_0 + b_1 x$$

where $b_1 = SS_{xy}/SS_{xx}$, $b_0 = \bar{y} - b_1\bar{x}$, $SS_{xx} = \sum(x - \bar{x})^2$ and $SS_{xy} = \sum(x - \bar{x})(y - \bar{y})$.

The Minitab pull-down **Stat ⇒ Regression ⇒ Fitted Line Plot** plots the data and the line that fits through the data.

The Excel pull-down **Tools ⇒ Data Analysis** produces the Data Analysis dialog box and from this box Regression is selected.

The *slope of the regression line* is tested by the following test statistic (H_0: $\beta_1 = c$ versus H_a: $\beta_1 \neq c$):

$$t = \frac{b_1 - c}{\text{standard error of } b_1}$$

and t has $n - 2$ degrees of freedom.

The *correlation coefficient r* may be found by using Minitab or Excel. The formula for r is

$$r = \frac{SS_{xy}}{\sqrt{SS_{xx}SS_{yy}}}$$

The x and y values are put into columns C1 and C2 of the Minitab worksheet and the pull-down **Stat ⇒ Basic Statistics ⇒ Correlation** is performed. If Excel is used to compute the correlation coefficient, the pull-down **Tools ⇒ Data Analysis** is used and Correlation is selected from the Data Analysis dialog box.

The square of the correlation coefficient is called the *coefficient of determination*.

A $(1 - \alpha)100\%$ *prediction interval* for an individual new value of y at $x = x_0$ is

$$\hat{y} \pm t_{\alpha/2}\, s \sqrt{1 + \frac{1}{n} + \frac{(x_0 - \bar{x})^2}{SS_{xx}}}$$

where $\hat{y} = b_0 + b_1 x_0$, the t value is based on $(n - 2)$ degrees of freedom, $s = \sqrt{SSE/(n - 2)}$ and is referred to as the estimated standard error of the regression model, n is the sample size, x_0 is the fixed value of x, $\bar{x}$ is the mean of the observed x values, and $SS_{xx} = \sum(x - \bar{x})^2$.

A $(1 - \alpha)100\%$ *confidence interval* for the mean value of y at $x = x_0$ is

$$\hat{y} \pm t_{\alpha/2}\, s \sqrt{\frac{1}{n} + \frac{(x_0 - \bar{x})^2}{SS_{xx}}}$$

and the parts of the equation are as defined in the prediction interval discussion. The prediction interval and the confidence interval may be found using Minitab. The Minitab pull-down **Stat** ⇒ **Regression** ⇒ **Regression** gives the Regression dialog box. Using the options in this dialog box, you may request a prediction interval and a confidence interval at any level of confidence.

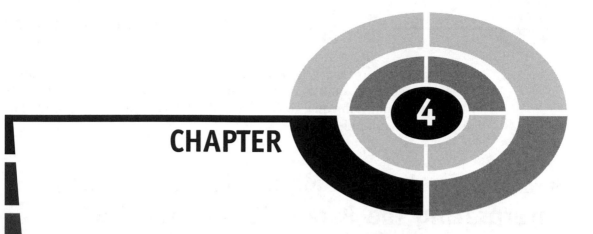

CHAPTER 4

Multiple Regression

4-1 Multiple Regression Models

Rather than try to model a dependent variable by a single independent variable as is done in Chapter 3, sometimes we model the dependent variable by several independent variables. We may try to explain blood pressure not only by weight, but by age as well. We know that blood pressure tends to increase as weight increases but also increases as we age. To give another example, the price of a home may be modeled as a function of the number of baths, the number of bedrooms, the size of the lot, and the total square footage of a house.

In general, if y is the dependent variable and $x_1, x_2, \ldots, x_k$ are k independent variables, then the general multiple regression model has the general form

$$y = \beta_0 + \beta_1 x_1 + \beta_2 x_2 + \cdots + \beta_k x_k + \varepsilon$$

The part $E(y) = \beta_0 + \beta_1 x_1 + \beta_2 x_2 + \cdots + \beta_k x_k$ is the deterministic portion of the model. The ε term is the random error term.

The *assumptions of multiple regression* are: (1) For any given set of values of the independent variables, the random error ε has a normal probability distribution with mean equal to 0 and standard deviation equal to σ. (2) The random errors are independent.

4-2 The First-Order Model: Estimating and Interpreting the Parameters in the Model

Purpose: The purpose of this section is to fit a regression model to a set of data and to determine the values of the coefficients. The interpretation of the coefficients is also discussed.

Assumptions: The assumptions of multiple regression.

EXAMPLE 4-1
A real-estate executive would like to be able to predict the cost of a house in a housing development on the basis of the number of bedrooms and bathrooms in the house. Table 4-1 contains information on the selling price of homes (in thousands of dollars), the number of bedrooms, and the number of baths.

The following first-order model is assumed to connect the selling price of the home with the number of bedrooms and the number of baths. The dependent variable is represented by y and the independent variables are x_1, the number of bedrooms, and x_2, the number of baths.

$$y = \beta_0 + \beta_1 x_1 + \beta_2 x_2 + \varepsilon$$

Calculus techniques are used to estimate the values of β_0, β_1, and β_2 so that this model fits the data in the table best in the least-squares sense. Both Minitab and Excel incorporate these techniques and give the following estimated regression equation:

$$y = b_0 + b_1 x_1 + b_2 x_2$$

SOLUTION
When Minitab is used to find the estimated regression equation, the first step is to enter the data in columns C1, C2, and C3 of the worksheet and to name the columns price, bedrooms, and baths. This is shown in Fig. 4-1.

The pull-down **Stat $\Rightarrow$ Regression $\Rightarrow$ Regression** gives the dialog box shown in Fig. 4-2. When the OK box is clicked, the following output is produced.

Table 4-1   The price of houses is dependent on the number of bedrooms and baths.

Price	Bedrooms	Baths
154	3	3
176	4	3
223	4	4
160	3	4
242	5	3
230	5	4
259	5	5
227	4	5
164	4	3
231	5	5

↓	C1	C2	C3	C4
	Price	Bedrooms	Baths	
1	154	3	3	
2	176	4	3	
3	223	4	4	
4	160	3	4	
5	242	5	3	
6	230	5	4	
7	259	5	5	
8	227	4	5	
9	164	4	3	
10	231	5	5	
11				

Fig. 4-1.

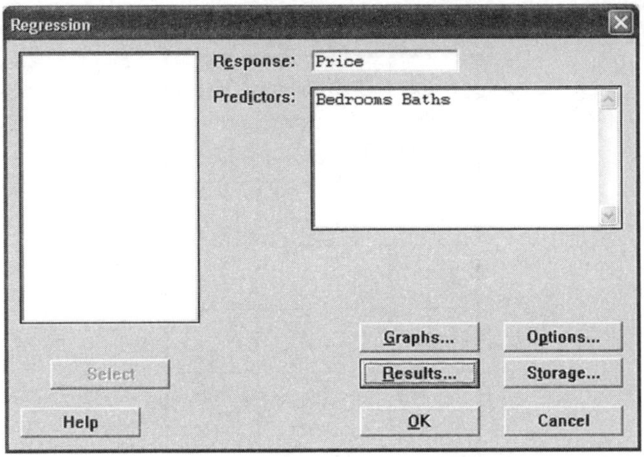

Fig. 4-2.

```
The regression equation is
price = - 5.0 + 35.7 bedrooms + 15.8 baths

Predictor    Coef        SE Coef    T        P
Constant     -5.05       35.56      -0.14    0.891
bedrooms     35.722      7.946      4.50     0.003
baths        15.799      7.158      2.21     0.063
```

The values for b_0, b_1, and b_2 are shown in the output ($b_0 = -5.05$, $b_1 = 35.722$, $b_2 = 15.799$). The estimated regression equation is

$$\text{Price} = -5.0 + 35.7 \text{ bedrooms} + 15.8 \text{ baths}$$

The estimated regression equation tells us that adding one additional bedroom adds $35,700 onto the price of the house and that adding on an additional bath will increase the price of the house by $15,800. The -5.0 is not interpretable in a practical sense.

A house for sale, from the housing development, with four bedrooms and two baths might be listed at

$$-5.0 + 35.7(4) + 15.8(2) = \$169,400$$

EXAMPLE 4-2
Solve Example 4-1 using Excel.

SOLUTION
The Excel solution to the same problem would proceed in the following manner. Data is entered into the Excel worksheet as shown in Fig. 4-3.

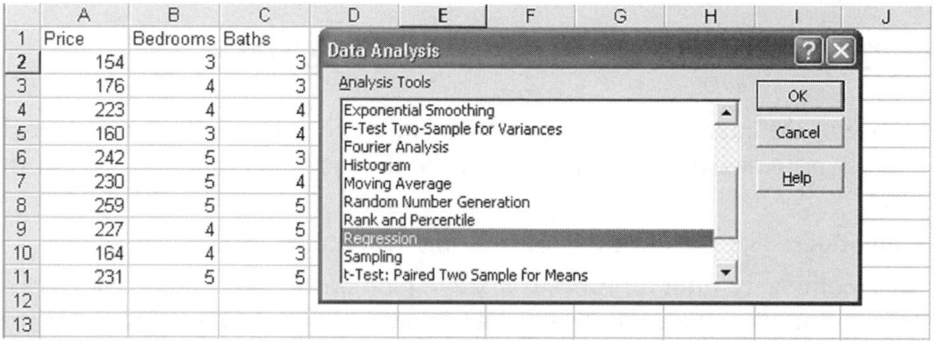

	A	B	C	D	E	F	G	H	I	J
1	Price	Bedrooms	Baths							
2	154	3	3							
3	176	4	3							
4	223	4	4							
5	160	3	4							
6	242	5	3							
7	230	5	4							
8	259	5	5							
9	227	4	5							
10	164	4	3							
11	231	5	5							
12										
13										

Data Analysis dialog box:

Analysis Tools
- Exponential Smoothing
- F-Test Two-Sample for Variances
- Fourier Analysis
- Histogram
- Moving Average
- Random Number Generation
- Rank and Percentile
- Regression
- Sampling
- t-Test: Paired Two Sample for Means

OK   Cancel   Help

Fig. 4-3.

The pull-down **Tools ⇒ Data Analysis** gives the data analysis dialog box and the Regression routine is selected.

The Excel output is shown in Fig. 4-4 and the values for the coefficients are shown.

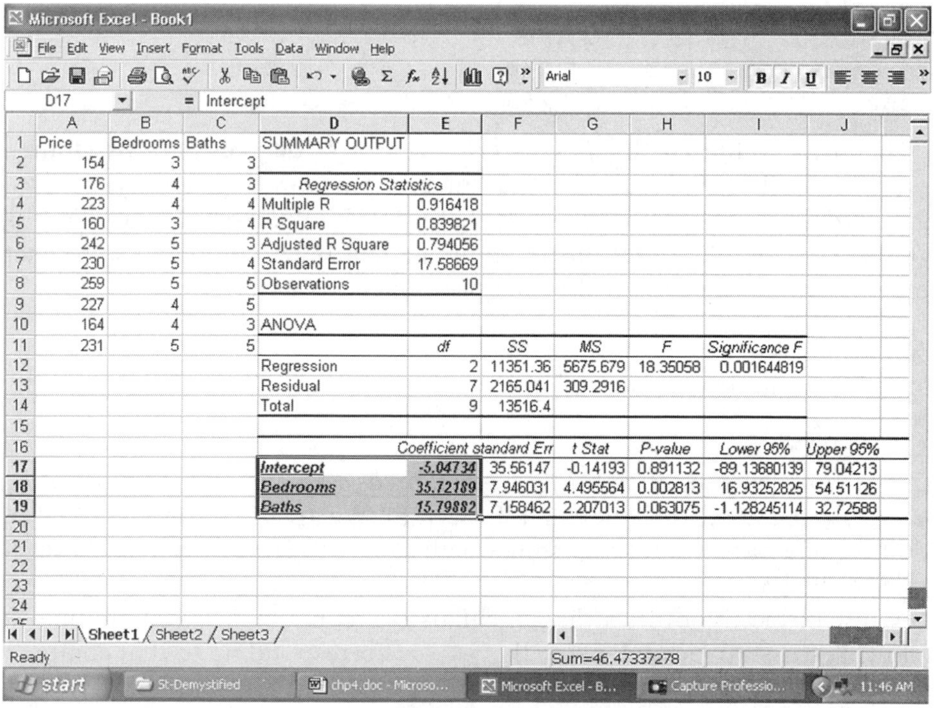

Microsoft Excel - Book1

File Edit View Insert Format Tools Data Window Help

D17   = Intercept

	A	B	C	D	E	F	G	H	I	J
1	Price	Bedrooms	Baths	SUMMARY OUTPUT						
2	154	3	3							
3	176	4	3	Regression Statistics						
4	223	4	4	Multiple R	0.916418					
5	160	3	4	R Square	0.839821					
6	242	5	3	Adjusted R Square	0.794056					
7	230	5	4	Standard Error	17.58669					
8	259	5	5	Observations	10					
9	227	4	5							
10	164	4	3	ANOVA						
11	231	5	5		df	SS	MS	F	Significance F	
12				Regression	2	11351.36	5675.679	18.35058	0.001644819	
13				Residual	7	2165.041	309.2916			
14				Total	9	13516.4				
15										
16					Coefficient	standard Err	t Stat	P-value	Lower 95%	Upper 95%
17				Intercept	-5.04734	35.56147	-0.14193	0.891132	-89.13680139	79.04213
18				Bedrooms	35.72189	7.946031	4.495564	0.002813	16.93252825	54.51126
19				Baths	15.79882	7.158462	2.207013	0.063075	-1.128245114	32.72588
20										
21										
22										
23										
24										

Sheet1 / Sheet2 / Sheet3 /

Ready   Sum=46.47337278

start   St-Demystified   chp4.doc - Microso...   Microsoft Excel - B...   Capture Professio...   11:46 AM

Fig. 4-4.

4-3 Inferences About the Parameters

Purpose of the inferences: The purpose of inferences about the parameters is to test hypotheses about the β_i or to set confidence intervals on the β_i in the model.

Assumptions: The assumptions of multiple regression.

The confidence interval for β_i is $b_i \pm t_{\alpha/2}$ (standard error b_i). The degrees of freedom for the t value is $n - (k+1)$, where $n =$ sample size, $(k+1) =$ the number of betas in the model.

EXAMPLE 4-3
Suppose we wished to set a 95% confidence interval on β_1, the coefficient in front of the bedrooms variable in Example 4-1.

SOLUTION
The standard errors of the b_i are shown in the following output. The standard error of b_1 is 7.946.

Predictor	Coef	SE Coef	T	P
Constant	−5.05	**35.56**	−0.14	0.891
bedrooms	35.722	**7.946**	4.50	0.003
baths	15.799	**7.158**	2.21	0.063

The $t_{.025}$ value with $n - (k+1) = 10 - 3 = 7$ degrees of freedom is $t_{.025} = 2.365$. The 95% confidence interval on β_1 is given by $b_1 \pm t_{.025}$ (standard error of b_1), which is $35.722 \pm 2.365(7.946)$ or $(16.930, 54.514)$. We are 95% confident that the addition of a bedroom would add somewhere between \$16,930 and \$54,514 to the price of the house.

EXAMPLE 4-4
Set a 95% confidence interval on β_2.

SOLUTION
A 95% confidence interval on β_2 is given by $b_2 \pm t_{.025}$ (standard error of b_2), i.e., $15.799 \pm 2.365(7.158)$ or $(−1.130, 32.728)$. We are 95% confident that the addition of a bath would add somewhere close to zero up to \$32,728 to the price of the house.

A test of the hypothesis that β_i equals c is conducted by computing the following test statistic and giving the p-value corresponding to that computed test statistic.

$$t = \frac{b_i - c}{\text{standard error of } b_i}$$

Note: The variable t has a t distribution with $n-(k+1)$ degrees of freedom.

> **Caution:** The reader is cautioned about conducting several t tests on the betas. If each t test is conducted at $\alpha = 0.05$, the actual alpha that would cover all the tests simultaneously is considerably larger than 0.05. For example, in testing the betas for significance (that is, $\beta_i = 0$), if tests are conducted on 10 betas, each at $\alpha = 0.05$, then if all the β parameters (except β_0) are equal to 0, approximately 40% of the time you will incorrectly reject the null hypothesis at least once and conclude that some β parameter differs from 0.

EXAMPLE 4-5

Table 4-2 contains data from a blood pressure study. The data were collected on a group of middle aged men. *Systolic* is the systolic blood pressure,

Table 4-2   Blood pressure study on fifty middle-aged men.

Systolic	Age	Weight	Parents	Med	Type A
141	46	207	0	6	18
153	47	215	0	1	23
137	36	190	0	2	17
139	46	210	2	1	16
135	44	214	2	5	18
139	45	224	1	9	10
133	45	237	2	9	9
150	56	229	1	7	18
131	45	179	0	1	15
133	50	215	0	2	15
146	53	217	0	10	18

(*Continued*)

Table 4-2   Continued.

Systolic	Age	Weight	Parents	Med	Type A
138	50	207	1	5	14
140	56	196	2	8	16
132	45	196	2	9	8
145	56	232	2	4	16
148	52	200	2	7	16
142	53	202	1	4	16
145	53	228	1	4	17
130	42	196	1	10	14
126	52	199	0	5	10
146	48	216	1	4	15
144	46	228	1	8	16
155	63	199	2	0	20
115	49	181	2	4	10
133	39	203	1	1	11
139	56	207	2	6	15
137	61	218	2	1	15
142	51	226	1	8	16
141	53	207	2	1	19
137	52	200	2	6	16

(Continued)

Table 4-2 Continued.

Systolic	Age	Weight	Parents	Med	Type A
141	54	194	1	6	16
131	40	206	1	5	16
147	47	221	0	1	17
134	53	200	0	1	17
144	57	209	0	3	17
140	47	212	0	10	17
132	54	177	0	10	17
138	48	202	0	2	12
115	46	185	0	6	12
137	50	199	2	0	15
151	53	229	2	0	18
145	54	203	2	3	14
139	61	207	2	1	14
144	43	215	1	9	15
126	53	214	1	0	11
134	44	195	1	0	13
158	43	238	2	10	14
116	43	195	2	5	8
141	56	209	1	7	15
139	52	222	1	10	13

Age is the age of the individual, *Weight* is the weight in pounds, *Parents* indicates whether the individual's parents had high blood pressure: 0 means neither parent has high blood pressure, 1 means one parent has high blood pressure, and 2 means both mother and father have high blood pressure, *Med* is the number of hours per month that the individual meditates, and *TypeA* is a measure of the degree to which the individual exhibits type A personality behavior, as determined from a form that the person fills out. Systolic is the dependent variable and the other five variables are the independent variables. The model assumed is

$$y = \beta_0 + \beta_1 x_1 + \beta_2 x_2 + \beta_3 x_3 + \beta_4 x_4 + \beta_5 x_5 + \varepsilon$$

where y = systolic, x_1 = age, x_2 = weight, x_3 = parents, x_4 = med, and x_5 = typeA.

SOLUTION
The data is entered into the Minitab worksheet. Part of the results is as follows.

```
The regression equation is
Systolic = 46.0 + 0.118 Age + 0.278 Weight + 1.23 Parents + 0.176
           Med + 1.78 TypeA
```

Predictor	Coef	SE Coef	T	P
Constant	45.95	12.74	3.61	0.001
Age	0.1181	0.1464	*0.81*	*0.424*
Weight	0.27812	0.05662	*4.91*	*0.000*
Parents	1.232	1.041	*1.18*	*0.243*
Med	0.1762	0.2405	*0.73*	*0.468*
TypeA	1.7752	0.2849	*6.23*	*0.000*

Five t tests are performed and the results are as follows.

$$H_0: \beta_1 = 0 \qquad H_a: \beta_1 \neq 0 \qquad p\text{-value} = 0.424$$

$$H_0: \beta_2 = 0 \qquad H_a: \beta_2 \neq 0 \qquad p\text{-value} = 0.000$$

$$H_0: \beta_3 = 0 \qquad H_a: \beta_3 \neq 0 \qquad p\text{-value} = 0.243$$

$$H_0: \beta_4 = 0 \qquad H_a: \beta_4 \neq 0 \qquad p\text{-value} = 0.468$$

$$H_0: \beta_5 = 0 \qquad H_a: \beta_5 \neq 0 \qquad p\text{-value} = 0.000$$

The five t-tests suggest weight and typeA should be kept and the other three variables thrown out. If tests are conducted on five betas, each at $\alpha = 0.05$, then if all the β parameters (except β_0) are equal to 0, approximately 22.6% of the time you will incorrectly reject the null hypothesis at least once and conclude that some β parameter differs from 0.

4-4 Checking the Overall Utility of a Model

Purpose: The purpose is to check whether the model is useful and to control your α value.

Assumptions: The assumptions of multiple regression.

EXAMPLE 4-6

Rather than conduct a large group of t-tests on the betas and increase the probability of making a type I error, we prefer to make one test and know that $\alpha = 0.05$. The F-test is such a test. It is contained in the analysis of variance associated with the analysis. The F-test tests the following hypothesis associated with the blood pressure model in Example 4-5.

$$H_0: \beta_1 = \beta_2 = \beta_3 = \beta_4 = \beta_5 = 0 \quad \text{versus} \quad H_a: \text{At least one } \beta_i \neq 0$$

SOLUTION

The following is part of the Minitab analysis associated with describing the blood pressure of middle-aged men.

```
Analysis of Variance
Source              DF      SS        MS        F         P
Regression           5   2740.92    548.18    18.50     0.000
Residual Error      44   1303.56     29.63
Total               49   4044.48
```

As is seen, $F = 18.50$ with a p-value of 0.000 and the null hypothesis should be rejected; the conclusion is that at least one $\beta_i \neq 0$. This F-test says that the model is useful in predicting systolic blood pressure. The Excel output associated with the analysis of variance is shown in Fig. 4-5.

10 ANOVA					
11	df	SS	MS	F	Significance F
12 Regression	5	2740.917562	548.1835	**18.5032**	**7.37566E-10**
13 Residual	44	1303.562438	29.62642		
14 Total	49	4044.48			

Fig. 4-5.

The *coefficient of determination* is also an important measure. It is shown as *R-Sq* in the following Minitab output. The adjusted coefficient of determination has been adjusted to take into account the sample size and the number of independent variables. This is shown as *R-Sq(adj)* in the following output. *The coefficient of determination gives the percent of systolic blood pressure variation accounted for by the model.*

```
The regression equation is
Systolic = 46.0 + 0.118 Age + 0.278 Weight + 1.23 Parents + 0.176
           Med + 1.78 TypeA

Predictor    Coef       SE Coef     T        P
Constant     45.95      12.74       3.61     0.001
Age          0.1181     0.1464      0.81     0.424
Weight       0.27812    0.05662     4.91     0.000
Parents      1.232      1.041       1.18     0.243
Med          0.1762     0.2405      0.73     0.468
TypeA        1.7752     0.2849      6.23     0.000
```

$S = 5.443$ **R-Sq = 67.8%**   **R-Sq(adj) = 64.1%**

R-Sq(adj) is defined in terms of R-Sq as follows

$$\text{R-Sq(adj)} = 1 - \left[\frac{(n-1)}{n-(k+1)}\right](1 - \text{R-Sq})$$

In the blood pressure example, $n = 50$, $k = 5$, and R-Sq $= 67.8$.

$$1 - \left[\frac{(50-1)}{50-(5+1)}\right](1 - 0.678) = 0.641 \text{ or R-Sq(adj)} = 64.1\%.$$

The part of the Excel output that gives the R square and Adjusted R square is shown in Fig. 4-6.

3	Regression Statistics	
4	Multiple R	0.823221379
5	R Square	0.67769344
6	Adjusted R Square	0.64106769
7	Standard Error	5.44301562
8	Observations	50

Fig. 4-6.

4-5 Using the Model for Estimation and Prediction

Purpose: Using Minitab, we may obtain confidence intervals for means as well as obtain prediction intervals for individual predictions that we make with the estimated regression equation.

Assumptions: The assumptions of multiple regression.

EXAMPLE 4-7

Based on the data given in Table 4-3, the estimated regression equation was found to be

$$\text{Price} = -5.0 + 35.7\ \text{bedrooms} + 15.8\ \text{baths}$$

Table 4-3　Price as a function of number of bedrooms and baths.

Price	Bedrooms	Baths
154	3	3
176	4	3
223	4	4
160	3	4
242	5	3
230	5	4
259	5	5
227	4	5
164	4	3
231	5	5

Just as we did in simple regression, we can predict the price of a home that has three bedrooms and three baths or we may estimate the average price of all homes that have three bedrooms and three baths. In both cases the point estimate will be $\text{Price} = -5.0 + 35.7(3) + 15.8(3) = 149.5$ thousand or \$149,500. The Minitab options dialog box is filled in as shown in Fig. 4-7. This dialog box produces the following prediction and confidence intervals.

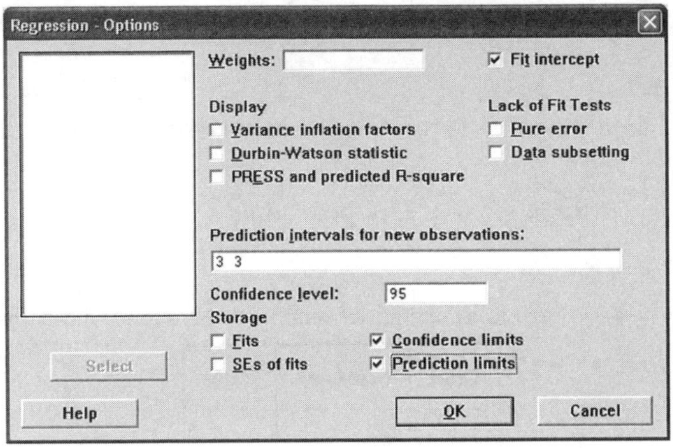

Fig. 4-7.

SOLUTION

Predicted Values for New Observations

New Obs	Fit	SE Fit	95.0% CI	95.0% PI
1	149.51	10.95	*(123.61, 175.42)*	*(100.50, 198.53)*

The 95% confidence interval for the mean price of all homes with three bedrooms and three baths in the sampled area is (123.61, 175.42) or $123,610 to $175,420. The 95% prediction interval for a home in the sampled area having three bedrooms and three baths is $100,500 to $198,530.

Excel is not programmed to compute confidence and prediction intervals.

4-6 Interaction Models

Purpose of including interaction terms in a model: If an experiment indicates interaction is present, then include the cross product term x_1x_2, in the linear model. Interaction is defined in the following paragraph.

Assumptions: The assumptions of multiple regression.

Suppose we have a model with two independent variables. If we anticipate that the response variable's relationship to x_1 may depend on the value of x_2, we might want to include a cross product, x_1x_2, in the model. This cross product is called an *interaction term*. For example, suppose we are studying the relationship between wheat yield and level of fertilizer and level of moisture. If we know that, at low levels of moisture, going from a low level to a high level of fertilizer will increase wheat yield whereas, at high levels of moisture, going from a low level to a high level of fertilizer will decrease the yield, then we may wish to model our wheat yield as $y = \beta_0 + \beta_1 x_1 + \beta_2 x_2 + \beta_3 x_1 x_2 + \varepsilon$. Consider the following example of wheat yield and moisture level and fertilizer level (Table 4-4). Eight plots of the same size have been selected, the moisture level and the fertilizer level controlled and the yield is measured on each plot.

EXAMPLE 4-8

Table 4-4   How the interaction of moisture level and fertilizer level affects wheat yield.

Plot	Wheat yield	Moisture	Fertilizer
1	22.5	10.1	1.6
2	22.7	10.2	1.5
3	33.4	10.4	3.5
4	33.3	9.8	3.7
5	25.5	22.1	1.6
6	26.0	22.5	1.5
7	17.2	22.3	3.7
8	16.9	22.4	3.6

Note that in plots 1 through 4 the moisture level is low and in plots 5 through 8 it is high. In plots 1 and 2 the fertilizer level is low and in plots 3 and 4 it is high. The wheat yield goes from low to high in these plots. In plots 5 and 6 the fertilizer level is low and in 7 and 8 the fertilizer level is high but the moisture level is high in these four plots and the yield goes down as the fertilizer level goes from low to high. This is an example of interaction. There is an interaction between moisture and fertilizer in their effect on wheat yield.

SOLUTION

Now consider fitting the first-order model without interaction, $y = \beta_0 + \beta_1 x_1 + \beta_2 x_2 + \varepsilon$, and the second-order model with interaction, $y = \beta_0 + \beta_1 x_1 + \beta_2 x_2 + \beta_3 x_1 x_2 + \varepsilon$. The Minitab analysis for the model with the interaction term is

Regression Analysis: y versus x1, x2, x1*x2

```
The regression equation is
y = 0.16 + 1.43 x1 + 12.8 x2 - 0.760 x1*x2
```

Predictor	Coef	SE Coef	T	P
Constant	0.160	2.035	0.08	0.941
x1	1.4314	0.1170	12.23	*0.000*
x2	12.8485	0.7297	17.61	*0.000*
x1*x2	−0.75967	0.04180	−18.17	*0.000*

S = 0.7520   R-Sq = 99.2%   *R-Sq(adj) = 98.6%*

Analysis of Variance

Source	DF	SS	MS	F	P
Regression	3	275.647	91.882	*162.49*	*0.000*
Residual Error	4	2.262	0.565		
Total	7	277.909			

$$H_0: \beta_1 = 0. \quad H_a: \beta_1 \neq 0. \quad p\text{-value} = 0.000$$
$$H_0: \beta_2 = 0. \quad H_a: \beta_2 \neq 0. \quad p\text{-value} = 0.000$$
$$H_0: \beta_3 = 0. \quad H_a: \beta_3 \neq 0. \quad p\text{-value} = 0.000$$

The three t-tests are all significant.

$$H_0: \beta_1 = \beta_2 = \beta_3 = 0 \quad \text{versus} \quad H_a: \text{At least one } \beta_i \neq 0$$

The F-test is significant.

The adjusted R-square is 98.6 %.

The Minitab analysis for the model without the interaction term is

Regression Analysis: y versus x1, x2

```
The regression equation is
y = 32.4 - 0.542 x1 + 0.43 x2
```

Predictor	Coef	SE Coef	T	P
Constant	32.377	8.171	3.96	0.011
x1	−0.5420	0.3562	−1.52	*0.189*
x2	0.427	2.091	0.20	*0.846*

S = 6.149   R-Sq = 32.0%   *R-Sq(adj) = 4.8%*

Analysis of Variance

Source	DF	SS	MS	F	P
Regression	2	88.87	44.44	**1.18**	**0.382**
Residual Error	5	189.04	37.81		
Total	7	277.91			

The three t-tests are all non-significant.

$$H_0: \beta_1 = \beta_2 = \beta_3 = 0 \quad \text{versus} \quad H_a: \text{At least one } \beta_i \neq 0$$

The F-test is non-significant and the adjusted R-square is 4.8%. Which model would you prefer?

4-7 Higher Order Models

Purpose: The purpose of higher order models is to account for curvature in the data.

Assumptions: The assumptions of multiple regression.

First consider the case when one independent variable is being considered. A plot of the dependent variable versus the independent variable may indicate a quadratic curvature to the data. In this case, the researcher may try a quadratic model that has the following form:

$$y = \beta_0 + \beta_1 x + \beta_2 x^2 + \varepsilon$$

The reader may recall from algebra that if $\beta_2 > 0$ the curve is concave upward and if $\beta_2 < 0$ then the curve is concave downward.

EXAMPLE 4-9

In a study of the growth of e-mail use versus time, suppose that the following data (Table 4-5) were collected at a three large companies over the past five years. The years have been coded as 1 through 5.

A plot of the data reveals a quadratic shape rather than a linear shape. The data in Table 4-5 is plotted in Fig. 4-8. The plot reveals that a quadratic model would be appropriate. The following model is assumed:

$$\text{e-mail} = \beta_0 + \beta_1 \text{ year} + \beta_2 \text{ yearsq} + \varepsilon$$

SOLUTION

In the Minitab worksheet, the number of e-mails is entered into C1, the coded years into C2, and the squares of the coded years into C3.

Table 4-5 E-mails have grown as a quadratic function.

Year (coded)	e-mails/employee
1	1.1
1	1.5
1	2.1
2	4.7
2	5.5
2	5.2
3	8.9
3	10.1
3	14.2
4	30.3
4	35.5
4	33.3
5	55.0
5	60.7
5	75.5

The fitted regression model is

Regression Analysis: e-mail versus year, yearsq

The regression equation is

e-mail = 12.4 - 14.9 year + 5.02 yearsq

Predictor	Coef	SE Coef	T	P
Constant	12.387	6.033	2.05	0.063
year	−14.905	4.597	−3.24	0.007
yearsq	5.0214	0.7518	6.68	0.000

S = 4.872   R-Sq = 96.6%   **R-Sq(adj) = 96.0%**

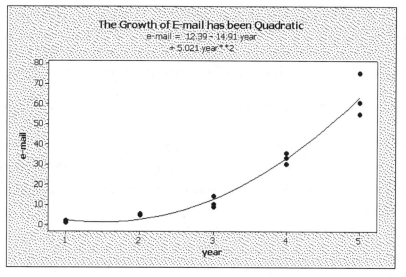

Fig. 4-8.

Analysis of Variance

Source	DF	SS	MS	F	P
Regression	2	8011.5	4005.8	**168.76**	**0.000**
Residual Error	12	284.8	23.7		
Total	14	8296.3			

The overall indicators show a good fit: $R\text{-Sq(adj)} = 96\%$ and $F = 168.76$ with corresponding p-value $= 0.000$.

If a linear model had been fit rather than a quadratic model the following results would have been obtained.

The regression equation is
e-mail = -22.8 + 15.2 year

Predictor	Coef	SE Coef	T	P
Constant	−22.763	6.157	−3.70	0.003
year	15.223	1.856	8.20	0.000

S = 10.17 $R\text{-Sq} = 83.8\%$ **R-Sq(adj) = 82.6%**

Analysis of Variance

Source	DF	SS	MS	F	P
Regression	1	6952.5	6952.5	**67.26**	**0.000**
Residual Error	13	1343.9	103.4		
Total	14	8296.3			

The R-Sq(adj) = 82.6%, the $F = 67.26$, which is smaller than that provided by the quadratic model. But notice that the standard error is reduced from $s = 10.17$ when the linear model is used to $s = 4.872$ when the quadratic model is used. Because of this reduction in standard error, better predictions will be made.

For year = 4, and the quadratic model, the prediction and confidence intervals are

```
New Obs     Fit       SE Fit     95.0% CI             95.0% PI
1           33.11     1.71       (29.37, 36.84)       (21.86, 44.36)
```

and for year = 4, and the linear model, the prediction and confidence intervals are

```
New Obs     Fit       SE Fit     95.0% CI             95.0% PI
1           38.13     3.22       (31.18, 45.08)       (15.09, 61.17)
```

The 95% CI for the linear model is (31.18, 45.08) and the 95% CI for the quadratic model is (29.37, 36.84). The 95% PI for the linear model is (15.09, 61.17) and the 95% PI for the quadratic model is (21.86, 44.36).

The Excel worksheet for this example is shown in Fig. 4-9.

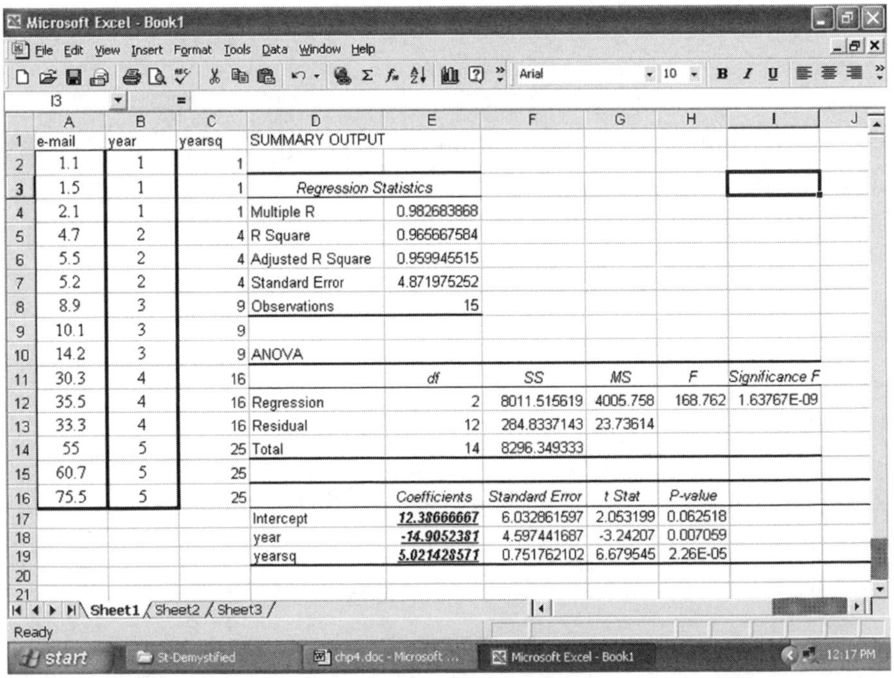

	A	B	C	D	E	F	G	H	I
1	e-mail	year	yearsq	SUMMARY OUTPUT					
2	1.1	1	1						
3	1.5	1	1	Regression Statistics					
4	2.1	1	1	Multiple R	0.982683868				
5	4.7	2	4	R Square	0.965667584				
6	5.5	2	4	Adjusted R Square	0.959945515				
7	5.2	2	4	Standard Error	4.871975252				
8	8.9	3	9	Observations	15				
9	10.1	3	9						
10	14.2	3	9	ANOVA					
11	30.3	4	16		df	SS	MS	F	Significance F
12	35.5	4	16	Regression	2	8011.515619	4005.758	168.762	1.63767E-09
13	33.3	4	16	Residual	12	284.8337143	23.73614		
14	55	5	25	Total	14	8296.349333			
15	60.7	5	25						
16	75.5	5	25		Coefficients	Standard Error	t Stat	P-value	
17				Intercept	12.38666667	6.032861597	2.053199	0.062518	
18				year	-14.9052381	4.597441687	-3.24207	0.007059	
19				yearsq	5.021428571	0.751762102	6.679545	2.26E-05	

Fig. 4-9.

When two independent variables are involved, a *complete second-degree model* is sometimes appropriate. A complete second-degree model in two independent variables is given as follows:

$$y = \beta_0 + \beta_1 x_1 + \beta_2 x_2 + \beta_3 x_1 x_2 + \beta_4 x_1^2 + \beta_5 x_2^2 + \varepsilon$$

The first term is the intercept term, the second and third terms are the linear terms, the fourth term is the interaction term, and the fifth and sixth terms are the second degree terms other than interaction.

4-8 Qualitative (Dummy) Variable Models

Purpose: One of the purposes of this section is to show the connection between regression and analysis of variance. Another purpose is the introduction of the concept of dummy variables.

Assumptions: The assumptions of multiple regression.

EXAMPLE 4-10
A group of fifteen diabetics with maturity onset diabetes was available for a research study. Five were randomly selected and assigned to an exercise group, five more were selected and assigned to a dietary group, and the remaining five were treated with insulin. After six months of treatment, their hemoglobin A1C values were determined. The lower the value of the hemoglobin A1C, the better the control of the diabetes. As you will recognize, this is a completely randomized design. It could be analyzed using the technique given in Chapter 2. However, we will analyze it using multiple regression and qualitative or dummy independent variables. The data are given in Table 4-6.

The one-way ANOVA is used to test H_0: $\mu_1 = \mu_2 = \mu_3$ versus the alternative that at least two of the means differ. (μ_1 is the mean A1C for the exercise treatment, μ_2 is the mean A1C for the dietary treatment, and μ_3 is the mean A1C for the insulin treatment.) We shall develop a multiple regression approach to testing the hypothesis.

SOLUTION
Define x_1 and x_2 as follows:

$$x_1 = \begin{cases} 1, \text{ if treatment } = \text{dietary} \\ 0, \text{ if treatment } = \text{other} \end{cases} \text{ and } x_2 = \begin{cases} 1, \text{ if treatment } = \text{insulin} \\ 0, \text{ if treatment } = \text{other} \end{cases}$$

When the independent variables are defined this way, they are called **dummy variables**.

Table 4-6   Hemoglobin A1C values of diabetics after six months.

Exercise	Dietary	Insulin
6.2	6.0	5.8
6.0	6.2	6.0
6.5	6.3	6.0
6.8	6.2	6.3
7.0	6.5	6.2

Now define the regression model as $y = \beta_0 + \beta_1 x_1 + \beta_2 x_2 + \varepsilon$. $E(y) = \beta_0 + \beta_1 x_1 + \beta_2 x_2$ is the deterministic part.

For treatment exercise, $E(y) = \beta_0 + \beta_1(0) + \beta_2(0) = \beta_0$ is the mean, or $\mu_1 = \beta_0$.
For treatment dietary, $E(y) = \beta_0 + \beta_1(1) + \beta_2(0) = \beta_0 + \beta_1$ is the mean, or $\mu_2 = \beta_0 + \beta_1$.
For treatment insulin, $E(y) = \beta_0 + \beta_1(0) + \beta_2(1) = \beta_0 + \beta_2$ is the mean, or $\mu_3 = \beta_0 + \beta_2$.

When the null hypothesis, $H_0: \beta_1 = \beta_2 = 0$ is true, $\mu_1 = \beta_0$, $\mu_2 = \beta_0 + \beta_1 = \beta_0 + 0 = \beta_0$, and $\mu_3 = \beta_0 + \beta_2 = \beta_0 + 0 = \beta_0$ or $\mu_1 = \mu_2 = \mu_3 = \beta_0$. Therefore, the global test that $H_0: \beta_1 = \beta_2 = 0$ is equivalent to the one-way ANOVA test.

The Minitab regression analysis proceeds as follows. The data is entered as shown in Fig. 4-10. Note that, when the values for x_1 and x_2 are filled in the Minitab worksheet, the values are obtained as given in the dummy variable definition above. The pull-down **Stat ⇒ Regression ⇒ Regression** produces the dialog box shown in Fig. 4-11, which is filled in as shown. The following output is produced.

Regression Analysis: y versus x1, x2

```
The regression equation is
y = 6.50 − 0.260 x1 − 0.440 x2

Predictor      Coef        SE Coef       T           P
Constant       6.5000      0.1268        51.28       0.000
x1            −0.2600      0.1793       −1.45        0.173
x2            −0.4400      0.1793       −2.45        0.030

S = 0.2834     R-Sq = 33.7%       R-Sq(adj) = 22.6%
```

↓	C1	C2	C3	C4
	y	x1	x2	
1	6.2	0	0	
2	6.0	0	0	
3	6.5	0	0	
4	6.8	0	0	
5	7.0	0	0	
6	6.0	1	0	
7	6.2	1	0	
8	6.3	1	0	
9	6.2	1	0	
10	6.5	1	0	
11	5.8	0	1	
12	6.0	0	1	
13	6.0	0	1	
14	6.3	0	1	
15	6.2	0	1	

Fig. 4-10.

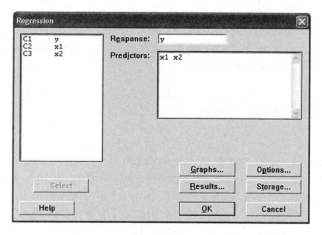

Fig. 4-11.

Analysis of Variance

Source	DF	SS	MS	F	P
Regression	2	0.48933	0.24467	*3.05*	*0.085*
Residual Error	12	0.96400	0.08033		
Total	14	1.45333			

The only part that we are interested in is underlined. It tells us not to reject H_0: $\beta_1 = \beta_2 = 0$ or not to reject the equality of the three means ($\mu_1 = \mu_2 = \mu_3 = \beta_0$) at $\alpha = 0.05$.

If the null hypothesis H_0: $\mu_1 = \mu_2 = \mu_3$ is tested using the techniques of Section 2.2, using Minitab, the following results:

One-way ANOVA: Exercise, Dietary, Insulin
Analysis of Variance

Source	DF	SS	MS	F	P
Factor	2	0.4893	0.2447	**3.05**	**0.085**
Error	12	0.9640	0.0803		
Total	14	1.4533			

```
                                        Individual 95% CIs For Mean
                                        Based on Pooled StDev
Level        N         Mean     StDev  - - + - - - + - - - + - - - -
Exercise     5         6.5000   0.4123                  (- -*- -)
Dietary      5         6.2400   0.1817            (- -*- -)
Insulin      5         6.0600   0.1949   (- -*- -)
                                         - - + - - - + - - - + - - - -
Pooled             StDev = 0.2834        6.00    6.30    6.60
```

The F value and the p-value are the same in the regression and the ANOVA outputs, showing the equivalence of the regression approach and the analysis of variance approach.

EXAMPLE 4-11
Solve Example 4-10 using Excel.

SOLUTION
The Excel solution to the regression problem is shown in Fig. 4-12. The Excel solution to the analysis of variance problem is shown in Fig. 4-13. Note once again the connection between the solution as a regression problem and the solution as an analysis of variance problem. The F value and the p-value are underlined in both Fig. 4-12 and Fig. 4-13.

EXAMPLE 4-12
To make sure the reader understands the technique involved in using dummy variables to test the equality of means, let's consider one more example. The number of dummy variables is always one less than the number of means that are being compared. Suppose commuting time in minutes is being compared for four cities. Five commuters are randomly selected from each city. The data are shown in Table 4-7.

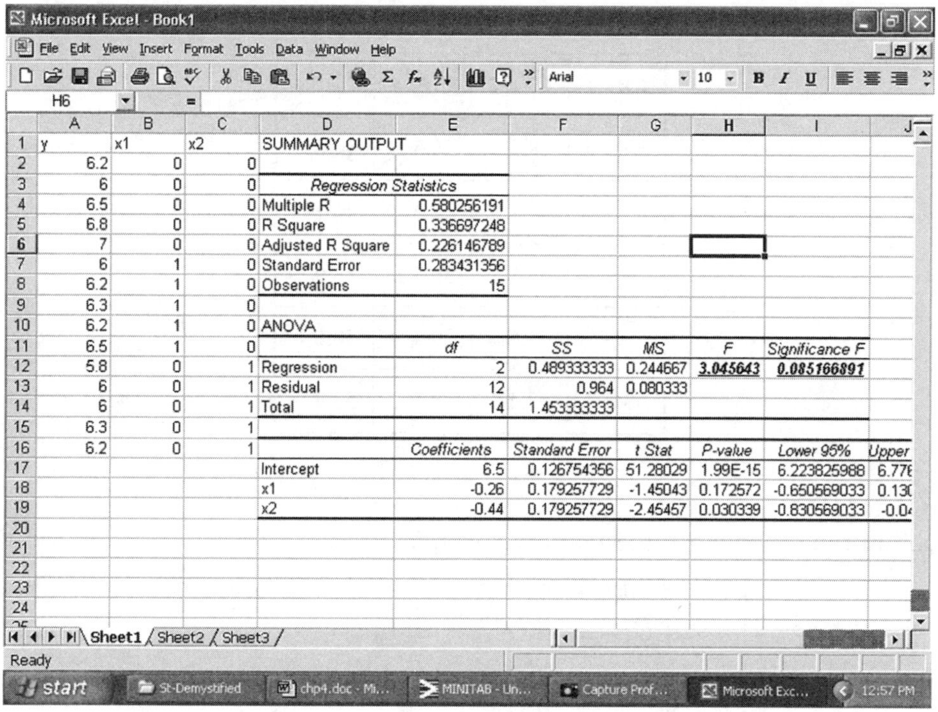

	A	B	C	D	E	F	G	H	I	J
1	y	x1	x2	SUMMARY OUTPUT						
2	6.2	0	0							
3	6	0	0	*Regression Statistics*						
4	6.5	0	0	Multiple R	0.580256191					
5	6.8	0	0	R Square	0.336697248					
6	7	0	0	Adjusted R Square	0.226146789					
7	6	1	0	Standard Error	0.283431356					
8	6.2	1	0	Observations	15					
9	6.3	1	0							
10	6.2	1	0	ANOVA						
11	6.5	1	0		*df*	*SS*	*MS*	*F*	*Significance F*	
12	5.8	0	1	Regression	2	0.489333333	0.244667	**3.045643**	**0.085166891**	
13	6	0	1	Residual	12	0.964	0.080333			
14	6	0	1	Total	14	1.453333333				
15	6.3	0	1							
16	6.2	0	1		*Coefficients*	*Standard Error*	*t Stat*	*P-value*	*Lower 95%*	*Upper*
17				Intercept	6.5	0.126754356	51.28029	1.99E-15	6.223825988	6.77
18				x1	-0.26	0.179257729	-1.45043	0.172572	-0.650569033	0.13
19				x2	-0.44	0.179257729	-2.45457	0.030339	-0.830569033	-0.0
20										
21										
22										
23										
24										

Fig. 4-12.

	A	B	C	D	E	F	G	H	I	J
1	Exercise	Dietary	Insulin	Anova: Single Factor						
2	6.2	6	5.8							
3	6	6.2	6	SUMMARY						
4	6.5	6.3	6	*Groups*	*Count*	*Sum*	*Average*	*Variance*		
5	6.8	6.2	6.3	Exercise	5	32.5	6.5	0.17		
6	7	6.5	6.2	Dietary	5	31.2	6.24	0.033		
7				Insulin	5	30.3	6.06	0.038		
8										
9										
10				ANOVA						
11				*Source of Variation*	*SS*	*df*	*MS*	*F*	*P-value*	*F crit*
12				Between Groups	0.489333	2	0.244667	**3.045643**	**0.085167**	3.88529
13				Within Groups	0.964	12	0.080333			
14										
15				Total	1.453333	14				
16										

Fig. 4-13.

Table 4-7 Commuting times for commuters from four cities.

City 1	City 2	City 3	City 4
30.5	38.7	36.5	40.6
34.3	39.7	33.5	44.7
40.6	35.5	36.8	48.9
38.5	45.5	30.4	50.6
33.3	37.4	40.5	52.4

SOLUTION

Define the following three dummy variables:

$$x_1 = \begin{cases} 1, & \text{if city} = \text{city2} \\ 0, & \text{if city} = \text{other} \end{cases} \quad x_2 = \begin{cases} 1, & \text{if city} = \text{city3} \\ 0, & \text{if city} = \text{other} \end{cases}$$

$$x_3 = \begin{cases} 1, & \text{if city} = \text{city4} \\ 0, & \text{if city} = \text{other} \end{cases}$$

The Minitab worksheet is shown in Fig. 4-14.

↓	C1	C2	C3	C4
	y	x1	x2	x3
1	30.5	0	0	0
2	34.3	0	0	0
3	40.6	0	0	0
4	38.5	0	0	0
5	33.3	0	0	0
6	38.7	1	0	0
7	39.7	1	0	0
8	35.5	1	0	0
9	45.5	1	0	0
10	37.4	1	0	0
11	36.5	0	1	0
12	33.5	0	1	0
13	36.8	0	1	0
14	30.4	0	1	0
15	40.5	0	1	0
16	40.6	0	0	1
17	44.7	0	0	1

Fig. 4-14.

The regression equation is
y = 35.4 + 3.92 x1 + 0.10 x2 + 12.0 x3

Predictor	Coef	SE Coef	T	P
Constant	35.440	1.844	19.22	0.000
x1	3.920	2.608	1.50	0.152
x2	0.100	2.608	0.04	0.970
x3	12.000	2.608	4.60	0.000

S = 4.123   R-Sq = 63.6%   R-Sq(adj) = 56.8%

Analysis of Variance

Source	DF	SS	MS	F	P
Regression	3	476.08	158.69	*9.34*	*0.001*
Residual Error	16	271.97	17.00		
Total	19	748.05			

The p-value indicates that H_0: $\beta_1 = \beta_2 = \beta_3 = 0$ should be rejected, which implies that all four means are not equal. The reader should treat the problem as a one-way ANOVA problem and see that the same answer is obtained.

4-9 Models with Both Qualitative and Quantitative Variables

Purpose of studying such models: Many real-world situations have both types of variables and so we need to study models containing both. In addition we study how the use of both types can lead to model building in the real world.

Assumptions: The assumptions of multiple regression.

EXAMPLE 4-13
Consider a study concerning the effect of a qualitative variable as well as a quantitative variable on a response. The qualitative factor was gender, x_1, defined as follows:

$$x_1 = \begin{cases} 0, & \text{if male} \\ 1, & \text{if female} \end{cases}$$

and the quantitative factor was hours spent on the Internet per week, x_2.
    The response variable was cumulative GPA. The subjects were college seniors. The data were as shown in Table 4-8.

Table 4-8 GPA as a function of gender and time spent on the internet.

x_1	x_2		
	5	**10**	**15**
Male	2.5, 2.6, 2.7	3.0, 3.1, 3.2	3.5, 3.6, 3.5
Female	3.0, 3.1, 3.2	3.5, 3.6, 3.7	3.9, 3.9, 3.8

SOLUTION

The model assumed was $y = \beta_0 + \beta_1 x_1 + \beta_2 x_2 + \beta_3 x_1 x_2 + \varepsilon$. The data were entered into the Minitab worksheet as shown in Fig. 4-15. A portion of the output is shown as follows.

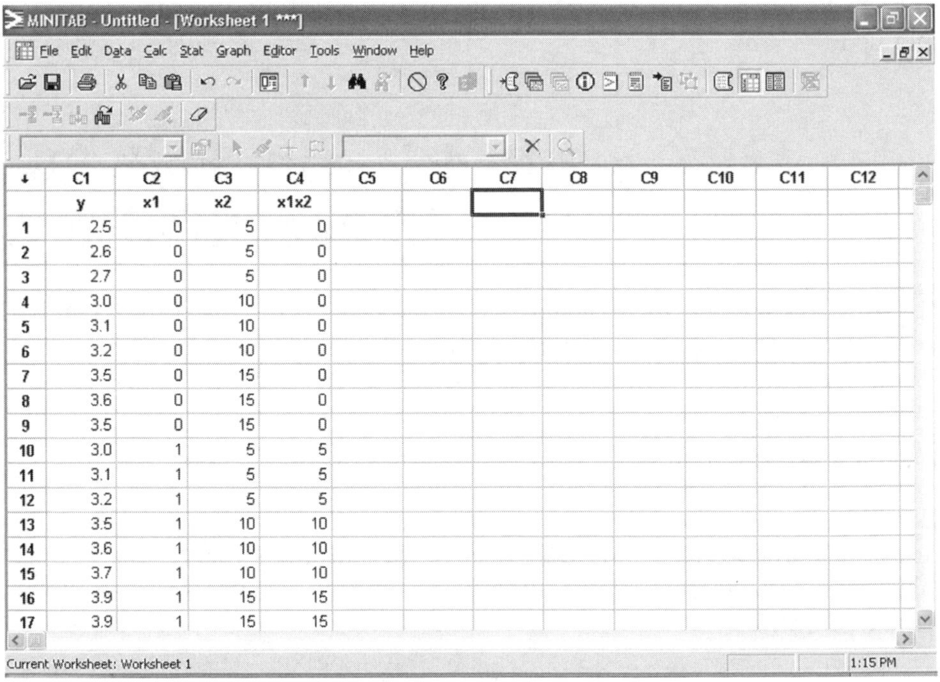

↓	C1	C2	C3	C4	C5	C6	C7	C8	C9	C10	C11	C12
	y	x1	x2	x1x2								
1	2.5	0	5	0								
2	2.6	0	5	0								
3	2.7	0	5	0								
4	3.0	0	10	0								
5	3.1	0	10	0								
6	3.2	0	10	0								
7	3.5	0	15	0								
8	3.6	0	15	0								
9	3.5	0	15	0								
10	3.0	1	5	5								
11	3.1	1	5	5								
12	3.2	1	5	5								
13	3.5	1	10	10								
14	3.6	1	10	10								
15	3.7	1	10	10								
16	3.9	1	15	15								
17	3.9	1	15	15								

Fig. 4-15.

```
The regression equation is
y = 2.14 + 0.611 x1 + 0.0933 x2 - 0.0167 x1x2

Predictor    Coef         SE Coef       T        P
Constant     2.14444      0.08259       25.97    0.000
x1           0.6111       0.1168        5.23     0.000
x2           0.093333     0.007646      12.21    0.000
x1x2        - 0.01667     0.01081      - 1.54    0.146
```

The test for interaction is underlined and is non-significant, since the p-value is 0.146.

The interaction plot is shown in Fig. 4-16. The solid line gives the male response and the dotted line gives the female response. The plot shows that

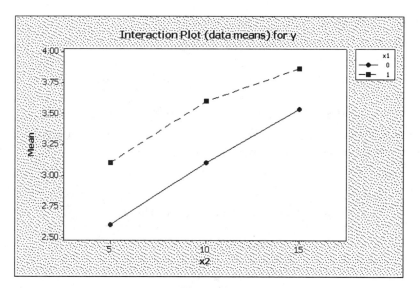

Fig. 4-16.

GPA increases as hours spent per week on the Internet increase. The female GPAs exceed the male GPAs by about 0.5 for similar times spent on the Internet. There is no interaction indicated for this data. The regression model is: $y = 2.14 + 0.611\ x_1 + 0.0933\ x_2 - 0.0167\ x_1x_2$. The equation of the male response is obtained by letting $x_1 = 0$. We get $y = 2.14 + 0.611(0) + 0.0933\ x_2 - 0.0167\ (0)x_2$ or $y = 2.14 + 0.0933x_2$. The female response is obtained by letting $x_1 = 1$. We get $y = 2.14 + 0.611(1) + 0.0933x_2 - 0.0167(1)x_2$ or $y = 2.751 + 0.0766x_2$.

Now, suppose the data were as shown in Table 4-9. Consider again the following portion of the Minitab output.

Table 4-9   A second set of data.

x_1	x_2		
	5	10	15
Male	2.5, 2.6, 2.7	3.0, 3.1, 3.2	2.7, 2.8, 2.8
Female	3.0, 3.1, 3.2	3.5, 3.6, 3.7	3.9, 3.9, 3.8

```
Predictor    Coef       SE Coef      T          P
Constant     2.6556     0.1612       16.47      0.000
x1           0.1000     0.2280        0.44      0.668
x2           0.01667    0.01492       1.12      0.283
x1x2         0.06000    0.02111       2.84      0.013
```

The underlined part shows that interaction is important with this data set. The new interaction plot is shown in Fig. 4-17.

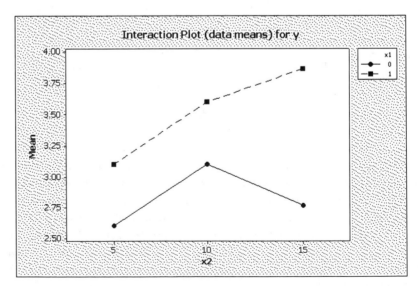

Fig. 4-17.

Now, this plot shows that too much time spent on the Internet by the males results in a decrease of the GPA. Perhaps the extra time spent is not related to the male studies but instead to other topics on the Internet.

Suppose a plot of the GPAs for males and females showed a quadratic relationship between y and x_2. This would suggest the following model:

$$y = \beta_0 + \beta_1 x_1 + \beta_2 x_2 + \beta_3 x_2^2 + \beta_3 x_1 x_2 + \beta_4 x_1 x_2^2 + \varepsilon$$

This model allows for a quadratic relationship between y and x_2 as well as interaction between the qualitative and quantitative terms.

4-10 Comparing Nested Models

Purpose of using reduced and complete models: This test allows us to determine whether certain terms should be retained in the model.

Assumptions: The assumptions of multiple regression.

EXAMPLE 4-14
Consider the second set of data from Example 4-13. The data are repeated in Table 4-10.

Table 4-10

x_1	x_2		
	5	**10**	**15**
Male	2.5, 2.6, 2.7	3.0, 3.1, 3.2	2.7, 2.8, 2.8
Female	3.0, 3.1, 3.2	3.5, 3.6, 3.7	3.9, 3.9, 3.8

The second degree model in two independent variables is $y = \beta_0 + \beta_1 x_1 + \beta_2 x_2 + \beta_3 x_2^2 + \beta_4 x_1 x_2 + \beta_5 x_1 x_2^2 + \varepsilon$. Suppose we wish to test that there is no quadratic part to the model; that is, we wish to test H_0: $\beta_3 = \beta_5 = 0$. This hypothesis, if true, would give a *reduced model* as follows:

$$y = \beta_0 + \beta_1 x_1 + \beta_2 x_2 + \beta_4 x_1 x_2 + \varepsilon$$

CHAPTER 4 **Multiple Regression**

The original model, $y = \beta_0 + \beta_1 x_1 + \beta_2 x_2 + \beta_3 x_2^2 + \beta_4 x_1 x_2 + \beta_5 x_1 x_2^2 + \varepsilon$, is called the *complete* model in two independent variables. We say that the reduced model is *nested* in the complete model.

SOLUTION

The worksheet set up for computing the complete model is shown in Fig. 4-18. (**Note:** the Minitab square of x_2 is C3**2 and is performed using

Fig. 4-18.

the Minitab calculator. The pull-down **Calc** $\Rightarrow$ **Calculator** activates the calculator.)

The complete model is fit to the data and the ANOVA part of the output is

Analysis of Variance

Source	DF	SS	MS	F	P
Regression	5	3.50278	0.70056	90.07	0.000
Residual Error	12	*0.09333*	*0.00778*		
Total	17	3.59611			

$SSE_C = 0.09333$ and $MSE_C = 0.00778$ are noted.

Next, the reduced model is fit to the data and the ANOVA part of the output is

```
Analysis of Variance
```

Source	DF	SS	MS	F	P
Regression	3	3.1283	1.0428	31.21	0.000
Residual Error	14	**0.4678**	**0.0334**		
Total	17	3.5961			

$SSE_R = 0.4678$

The test statistic for testing H_0: $\beta_3 = \beta_5 = 0$ is $F = [(SSE_R - SSE_C)/\# \text{ of } \beta s$ tested in $H_0]/MSE_C$ or $F = [(0.4678 - 0.09333)/2]/0.00778) = 24.1$. This F has 2 and 12 degrees of freedom. The p-value may be found as follows. Use the pull-down **Calc** $\Rightarrow$ **probability distribution** $\Rightarrow$ **F**. Fill in the F distribution dialog box as shown in Fig. 4-19. The following output is found.

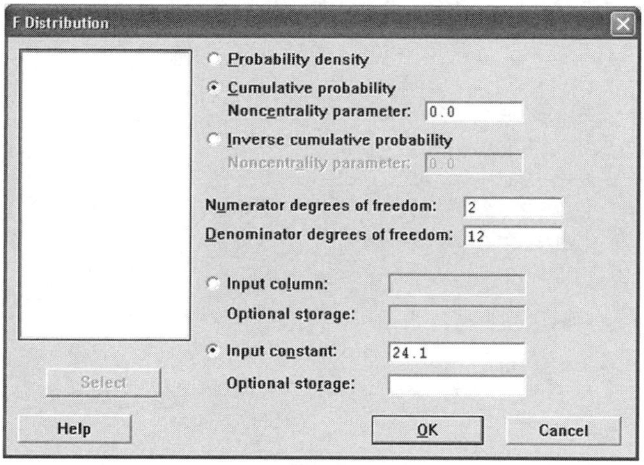

Fig. 4-19.

Cumulative Distribution Function

```
F distribution with 2 DF in numerator and 12 DF in denominator

x                 P(X<=x)
24.1              0.999937
```

The p-value is therefore $1 - 0.999937 = 0.000063$. This value is less than $\alpha = 0.05$. It appears that the quadratic terms are helpful in modeling the data.

EXAMPLE 4-15
As a second example, consider the data from Table 4-2 in Section 4-3. The complete first degree model is

$$y = \beta_0 + \beta_1 x_1 + \beta_2 x_2 + \beta_3 x_3 + \beta_4 x_4 + \beta_5 x_5 + \varepsilon$$

where $y =$ systolic, $x_1 =$ age, $x_2 =$ weight, $x_3 =$ parents, $x_4 =$ med, and $x_5 =$ typeA.

Suppose we wish to test the hypothesis H_0: $\beta_1 = \beta_3 = \beta_4 = 0$. The reduced model is

$$y = \beta_0 + \beta_2 x_2 + \beta_5 x_5 + \varepsilon$$

SOLUTION
The regression analysis for the complete model is

Regression Analysis: Systolic versus Age, Weight, Parents, Med, TypeA
```
The regression equation is
Systolic =  46.0 +  0.118 Age +  0.278 Weight +  1.23 Parents
            + 0.176 Med + 1.78 TypeA
```

Predictor	Coef	SE Coef	T	P
Constant	45.95	12.74	3.61	0.001
Age	0.1181	0.1464	0.81	0.424
Weight	0.27812	0.05662	4.91	0.000
Parents	1.232	1.041	1.18	0.243
Med	0.1762	0.2405	0.73	0.468
TypeA	1.7752	0.2849	6.23	0.000

```
S = 5.443    R-Sq = 67.8%    R-Sq(adj) = 64.1%
```

Analysis of Variance

Source	DF	SS	MS	F	P
Regression	5	2740.92	548.18	18.50	0.000
Residual Error	44	*1303.56*	*29.63*		
Total	49	4044.48			

$SSE_C = 1303.56$ and $MSE_C = 29.63$ are noted.
The regression analysis for the reduced model is

Regression Analysis: Systolic versus Weight, TypeA
```
The regression equation is
Systolic = 49.8 + 0.302 Weight + 1.73 TypeA
```

```
Predictor    Coef        SE Coef      T        P
Constant     49.75       11.39        4.37     0.000
Weight        0.30220     0.05415     5.58     0.000
TypeA         1.7254      0.2568      6.72     0.000

S=5.447     R-Sq=65.5%     R-Sq(adj)=64.1%
```

Analysis of Variance

```
Source             DF     SS        MS        F        P
Regression          2     2650.0    1325.0    44.66    0.000
Residual Error     47     1394.5    29.7
Total              49     4044.5
```

$SSE_R = 1394.5$

The test statistic for testing H_0: $\beta_1 = \beta_3 = \beta_4 = 0$ is

$$F = \frac{(SSE_R - SSE_C)/\# \text{ of } \beta\text{s tested in } H_0}{MSE_C} = \frac{(1394.5 - 1303.56)/3}{29.63} = 1.023$$

The p-value is computed as follows.

Cumulative Distribution Function

F distribution with 3 DF in numerator and 44 DF in denominator

```
x              P(X <= x)
1.023          0.6084
```

p-value $= 1 - 0.608 = 0.392$.

At $\alpha = 0.05$, we do not reject the null hypothesis H_0: $\beta_1 = \beta_3 = \beta_4 = 0$. The reduced model is adequate for describing blood pressure in terms of weight and type A personality. The terms involving age, parents, and meditation can be dropped without any loss.

4-11 Stepwise Regression

Purpose: The purpose of stepwise regression is to reduce the set of independent variables down to the most important predictors. Then interaction terms and curvilinear effects can be considered after the most important predictors have been identified.

Assumptions: The assumptions of multiple regression.

EXAMPLE 4-16

Consider a study of factors affecting systolic blood pressure in men. Eight independent variables have been identified for the study. The data are shown in Table 4-11, and the variables are defined below. A stepwise regression is to be performed on the data to reduce the number of independent variables that will be used to predict and explain the behavior of the systolic readings.

The variables are:

Table 4-11

Systolic	Age	Weight	Parents	Med	TypeA	Smoke	Drink	Exercise
141	46	207	0	6	18	1	2	5
153	47	215	0	1	23	1	4	0
137	36	190	0	2	17	0	2	5
139	46	210	2	1	16	0	3	5
135	44	214	2	5	18	0	2	0
139	45	224	1	9	10	0	2	5
133	45	237	2	9	9	0	3	3
150	56	229	1	7	18	1	6	0
131	45	179	0	1	15	0	2	10
133	50	215	0	2	15	0	2	10
146	53	217	0	10	18	1	3	5
138	50	207	1	5	14	0	2	5
140	56	196	2	8	16	1	4	5
132	45	196	2	9	8	0	2	7
145	56	232	2	4	16	1	4	5

(*Continued*)

Table 4-11   Continued.

Systolic	Age	Weight	Parents	Med	TypeA	Smoke	Drink	Exercise
148	52	200	2	7	16	1	5	2
142	53	202	1	4	16	1	2	2
145	53	228	1	4	17	1	4	5
130	42	196	1	10	14	0	2	10
126	52	199	0	5	10	0	1	15
146	48	216	1	4	15	1	3	5
144	46	228	1	8	16	1	4	5
155	63	199	2	0	20	1	7	0
115	49	181	2	4	10	0	0	15
133	39	203	1	1	11	0	3	3
139	56	207	2	6	15	0	2	5
137	61	218	2	1	15	0	3	5
142	51	226	1	8	16	1	4	5
141	53	207	2	1	19	1	4	0
137	52	200	2	6	16	0	4	3
141	54	194	1	6	16	1	3	5
131	40	206	1	5	16	0	2	10
147	47	221	0	1	17	1	4	0
134	53	200	0	1	17	0	2	5
144	57	209	0	3	17	1	1	3

(*Continued*)

Table 4-11 Continued.

Systolic	Age	Weight	Parents	Med	TypeA	Smoke	Drink	Exercise
140	47	212	0	10	17	1	3	3
132	54	177	0	10	17	0	2	10
138	48	202	0	2	12	0	2	5
115	46	185	0	6	12	0	0	15
137	50	199	2	0	15	0	2	5
151	53	229	2	0	18	1	6	3
145	54	203	2	3	14	1	4	5
139	61	207	2	1	14	0	3	5
144	43	215	1	9	15	1	2	3
126	53	214	1	0	11	0	1	10
134	44	195	1	0	13	0	2	5
158	43	238	2	10	14	1	8	0
116	43	195	2	5	8	0	0	15
141	56	209	1	7	15	1	3	3
139	52	222	1	10	13	0	2	2

$y =$ Systolic blood pressure
$x_1 =$ Age
$x_2 =$ Weight in pounds
$x_3 =$ Parents $=$ dummy variable

$$= \begin{cases} 0, \text{ if neither parent has high blood pressure} \\ 1, \text{ if one parent has high blood pressure} \\ 2, \text{ if both parents have high blood pressure} \end{cases}$$

$x_4 = $ Med $=$ number of hours spent meditating per month

$x_5 = $ TypeA $=$ a measure of type A personality. The higher the score, the more type A.

$$x_6 = \text{Smoke} = \text{dummy variable} = \begin{cases} 0, \text{ does not smoke} \\ 1, \text{ does smoke} \end{cases}$$

$x_7 = $ Drink $=$ number of ounces of alcohol consumed per week.

$x_8 = $ Exercise $=$ number of hours spent per week doing exercise.

SOLUTION

The data are entered into columns C1 through C9. The pull-down sequence **Stat** $\Rightarrow$ **Regression** $\Rightarrow$ **Stepwise** gives the Stepwise Regression dialog box, which is filled in as shown in Fig. 4-20.

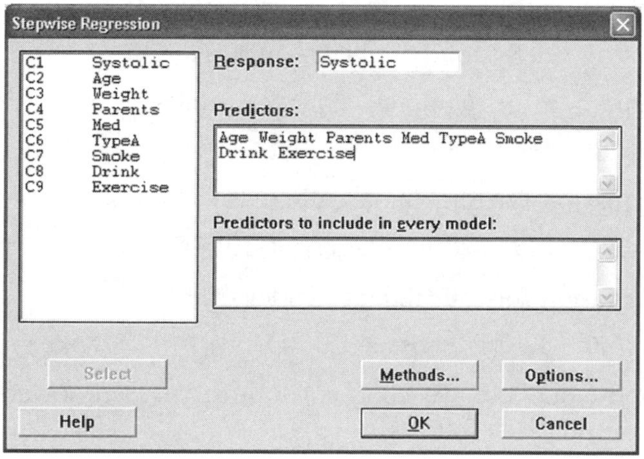

Fig. 4-20.

For Methods, choose forward selection and alpha to enter as 0.05. The output is as follows.

Stepwise Regression: Systolic versus Age, Weight,...

Forward selection. Alpha-to-Enter: 0.05 Response is Systolic on 8 predictors, with N$=50$

Step	1	2	3	4
Constant	148.4	136.8	135.8	136.6
Exercise	−1.90	−1.15	−1.02	−1.01
T-Value	−10.58	−6.11	−6.20	−6.48
P-Value	0.000	0.000	0.000	0.000

Drink		2.70	1.97	2.36
T-Value		5.83	4.57	5.32
P-Value		0.000	0.000	0.000
Smoke			5.1	4.4
T-Value			4.30	3.67
P-Value			0.000	0.001
Parents				−1.44
T-Value				−2.36
P-Value				0.023
S	5.03	3.87	3.30	3.15
R-Sq	69.99	82.59	87.59	88.95
R-Sq(adj)	69.36	81.85	86.78	87.97
C-p	77.0	27.4	8.9	5.3

The best equation for one independent variable is

$$\text{Systolic} = 148.4 - 1.90 \text{ Exercise}$$

The best equation for two independent variables is

$$\text{Systolic} = 136.8 - 1.15 \text{ Exercise} + 2.70 \text{ Drink}$$

The best equation for three independent variables is

$$\text{Systolic} = 135.8 - 1.02 \text{ Exercise} + 1.97 \text{ Drink} + 5.1 \text{ Smoke}$$

The best equation for four independent variables is

$$\text{Systolic} = 136.6 - 1.10 \text{ Exercise} + 2.36 \text{ Drink} + 4.4 \text{ Smoke} - 1.44 \text{ Parents}$$

For Methods, suppose we choose backward selection and enter alpha as 0.05. The output is as follows.

```
Backward elimination. Alpha-to-Remove: 0.05
Response is Systolic on 8 predictors, with N = 50
```

Step	1	2	3	4	5
Constant	116.2	116.2	121.5	131.7	136.6
Age	0.097	0.098	0.118	0.108	
T-Value	1.12	1.16	1.43	1.31	
P-Value	0.268	0.254	0.160	0.199	
Weight	0.060	0.060	0.047		
T-Value	1.51	1.53	1.26		
P-Value	0.140	0.135	0.213		
Parents	−1.37	−1.36	−1.66	−1.61	−1.44
T-Value	−1.98	−2.00	−2.69	−2.59	−2.36
P-Value	0.054	0.052	0.010	0.013	0.023

Med	-0.01				
T-Value	-0.06				
P-Value	0.950				
TypeA	0.23	0.23			
T-Value	0.98	1.04			
P-Value	0.334	0.306			
Smoke	3.5	3.5	3.8	4.0	4.4
T-Value	2.67	2.79	3.16	3.30	3.67
P-Value	0.011	0.008	0.003	0.002	0.001
Drink	2.13	2.13	2.24	2.34	2.36
T-Value	4.60	4.65	5.04	5.31	5.32
P-Value	0.000	0.000	0.000	0.000	0.000
Exercise	-0.89	-0.89	-0.97	-1.02	-1.01
T-Value	-5.01	-5.08	-6.08	-6.58	-6.48
P-Value	0.000	0.000	0.000	0.000	0.000
S	3.14	3.10	3.11	3.13	3.15
R-Sq	90.00	90.00	89.74	89.36	88.95
R-Sq(adj)	88.05	88.33	88.31	88.15	87.97
C-p	9.0	7.0	6.1	5.6	5.3

In the backward elimination, step 1 gives the equation with all eight independent variables included. Meditation is taken out in step 2. Meditation and type A are taken out in step 3. Meditation, type A, and weight are taken out in step 4. Meditation, type A, weight, and age are taken out in step 5. The equation that is given in step 5 is the same one that was given in the last step of the forward selection process. With four predictors in the regression equation, the value of R-Sq(adj) is close to 88%.

The C-p statistic is used in the evaluation of competing models. When a regression model with k independent variables contains only random differences from a true model, the average value of C-p is $k+1$. Thus, in evaluating several regression models, our objective is to find models whose C-p value is $\leq k+1$.

The C-p statistic is defined as follows:

$$C\text{-}p = \frac{(1 - R_k^2)(n - t)}{1 - R_t^2} - (n - 2(k + 1))$$

where
   $k =$ number of independent variables included in a regression model
   $t =$ total number of parameters (including the intercept) to be considered
        for inclusion in the regression model

$R_k^2 =$ coefficient of multiple determination for a regression model that has k independent variables

$R_t^2 =$ coefficient of multiple determination for a regression model that contains all t parameters

For example, in step 5 in the backward selection process above, the $C\text{-}p$ value is 5.3.

$$C\text{-}p = \frac{(1 - 0.8895)(50 - 9)}{1 - 0.90} - (50 - 10) = 5.3$$

4-12 Exercises for Chapter 4

1. An experiment was conducted and nine plots were available. The amount of fertilizer applied was 1, 2, or 3 units and the amount of moisture applied was 5, 10, or 15 units.

Table 4-12   Corn yield as a function of fertilizer and moisture.

Yield	Fertilizer	Moisture
10.4	1	5
18.8	1	10
24.9	1	15
10.2	2	5
17.9	2	10
26.7	2	15
13.8	3	5
22.1	3	10
33.9	3	15

(a) Fit the first-order linear model to the data in Table 4-12.
(b) Interpret b_0 in the fitted model $y = b_0 + b_1 x_1 + b_2 x_2$.

(c)   Interpret b_1 in the fitted model $y = b_0 + b_1x_1 + b_2x_2$.

(d)   Interpret b_2 in the fitted model $y = b_0 + b_1x_1 + b_2x_2$.

(e)   Give a point estimate of the yield when 2.5 units of fertilizer and 7.5 units of moisture are applied.

2.   Refer to the data in exercise 1.

(a)   Plot the main effects of fertilizer and moisture.

(b)   Plot the interaction graph.

(c)   Test for significant interaction, if possible.

3.   Use the data in exercise 1 to answer the following.

(a)   Fit the first-order model connecting yield to fertilizer and moisture. Test that fertilizer has a positive effect on yield. Test that moisture has a positive effect on yield.

(b)   Construct a 95% confidence interval on the average yield when the fertilizer applied is 2.5 units and the moisture applied is 7.5 units.

4.   The worldwide spam messages sent daily (in billions) is given in Table 4-13. Fit a quadratic model to the data and then use the model to predict the number to be sent daily in 2004. Assume the pattern of growth will continue.

Table 4-13   Spam as a quadratic function of time.

Year	Daily spam
1999	1.0
2000	2.3
2001	4.0
2002	5.6
2003	7.3

5.   The amount spent on medical expenses per year was related to other health factors for thirty adult males. A study collected the medical expenses per year, y, as well as information on the following

independent variables:

$$x_1 = \begin{cases} 0, \text{ if a non-smoker} \\ 1, \text{ if a smoker} \end{cases}$$

$x_2 =$ money spent on alcohol per week
$x_3 =$ hours spent exercising per week

$$x_4 = \begin{cases} 0, \text{ dietary knowledge is low} \\ 1, \text{ dietary knowledge is average} \\ 2, \text{ dietary knowledge is high} \end{cases}$$

$x_5 =$ weight
$x_6 =$ age.

Using the data in Table 4-14 and a first-order model, perform a forward stepwise regression.

Table 4-14 Medical cost as a function of six variables.

Medcost	Smoker	Alcohol	Exercise	Dietary	Weight	Age
2100	0	20	5	1	185	50
2378	1	25	0	1	200	42
1657	0	10	10	2	175	37
2584	1	20	5	2	225	54
2658	1	25	0	1	220	32
1842	0	0	10	1	165	34
2786	1	25	5	0	225	30
2178	0	10	10	1	180	41
3198	1	30	0	1	225	31
1782	0	5	10	0	180	45
2399	0	25	12	2	225	45
2423	0	15	15	0	220	33

(Continued)

Table 4-14   Continued.

Medcost	Smoker	Alcohol	Exercise	Dietary	Weight	Age
3700	1	25	0	1	275	43
2892	1	30	5	1	230	42
2350	1	30	10	1	245	40
2997	0	25	0	1	220	31
2678	0	20	25	0	245	39
2423	1	25	10	2	235	37
3316	1	35	5	0	250	31
2631	0	15	10	2	180	50
1860	1	20	15	0	220	49
2317	1	0	10	2	225	41
1870	1	25	15	1	220	34
1368	1	15	5	1	180	48
2916	0	10	10	2	200	46
1874	0	0	20	1	180	47
3739	0	15	15	2	280	31
2811	1	10	5	0	255	47
2912	0	15	10	1	210	45
2859	1	10	15	0	280	32

6. A study involved several companies involved in moving household goods. The response variable was the damage the company had to pay. The two independent variables were the weight of the goods and the distance traveled. The data were as given in Table 4-15.

Table 4-15 Damage payment as a function of distance moved and weight.

Damage, y	Distance, x_1	Weight, x_2
750	2300	4500
550	1000	3000
350	800	2500
800	1700	3800
975	2500	5000
1000	1500	4050
750	2800	3700
450	1250	3000
350	1760	2750
850	2400	1900

Fit the linear model with and without interaction. Comment on the fit of the two models.

7. Refer to problem 6. Use both models to find a 95% prediction interval and set a 95% confidence interval on damage when the trip is 2000 miles long and the weight is 4000 pounds. Compare the width of the confidence interval and the width of the prediction interval using the two models.

8. Using the data of problem 5, test that $H_0: \beta_2 = \beta_4 = \beta_6 = 0$ versus H_a: At least one of the three betas is not zero.

9. Using the data of problem 5, test that $H_0: \beta_1 = \beta_3 = \beta_5 = 0$ versus H_a: At least one of the three betas is not zero.

10. An experiment was conducted where $Y =$ corn production on similar plots, $X_1 =$ the fertilizer added to the plot, $X_2 =$ the moisture added to the plot, and $X_3 =$ the temperature of the plot. Analyze the results of the experiment. Assume a linear model and give your conclusions. The data is as given in Table 4-16.

Table 4-16   Corn yield as a function of three independent variables.

Y	X_1	X_2	X_3	Y	X_1	X_2	X_3
2	5	10	70	15	10	10	70
2	5	10	80	16	10	10	80
4	5	10	90	18	10	10	90
6	5	10	100	12	10	10	100
10	5	20	70	34	10	20	70
10	5	20	80	36	10	20	80
12	5	20	90	37	10	20	90
8	5	20	100	4	10	20	100

4-13 Chapter 4 Summary

If y is the dependent variable and $x_1, x_2, \ldots, x_k$ are k independent variables, then the general multiple regression first-order model has the general form

$$y = \beta_0 + \beta_1 x_1 + \beta_2 x_2 + \cdots + \beta_k x_k + \varepsilon$$

The beta coefficients are estimated from collected data and the estimated regression equation is represented by

$$\hat{y} = b_0 + b_1 x_1 + b_2 x_2 + \cdots + b_k x_k$$

The *assumptions of multiple regression* are: (1) For any given set of values of the independent variables, the random error ε has a normal probability distribution with mean equal to 0 and standard deviation equal to σ. (2) The random errors are independent.

The pull-down **Stat $\Rightarrow$ Regression $\Rightarrow$ Regression** is the Minitab command used to obtain the estimated regression equation and all the relevant regression output. The pull-down **Tools $\Rightarrow$ Data Analysis** is the Excel command used to obtain the estimated regression equation and all the relevant regression output.

The confidence interval for β_i is $b_i \pm t_{(\alpha/2)}$ (standard error b_i). The degrees of freedom for the t value is $n-(k+1)$, where $n =$ sample size and $(k+1) =$ the number of betas in the model.

A test of the hypothesis that β_i equals c is conducted by computing the following test statistic and giving the p-value corresponding to that computed test statistic.

$$t = \frac{b_i - c}{\text{standard error of } b_i}$$

Note: t has a t distribution with $n-(k+1)$ degrees of freedom.

The overall utility of a model is tested by the following hypothesis:

$$H_0: \ \beta_1 = \beta_2 = \cdots = \beta_k = 0 \quad \text{versus} \quad H_a: \text{At least one } \beta_i \neq 0$$

The F-test of the analysis of variance test is used to test this hypothesis. This is called the *global F-test*.

R-Sq(adj) is defined in terms of R-Sq as follows.

$$R\text{-Sq(adj)} = 1 - \left[\frac{(n-1)}{n-(k+1)}\right](1 - R\text{-Sq})$$

It is also a measure of the overall utility of the model.

The Regression Options dialog box can be used to set *prediction limits* or *confidence limits.*

When *interaction* is present, the model will require a cross product term of the form x_1x_2. A model with interaction will be of the form $y = \beta_0 + \beta_1x_1 + \beta_2x_2 + \beta_3x_1x_2 + \varepsilon$.

A *complete second-degree model in two independent variables* is given as follows:

$$y = \beta_0 + \beta_1x_1 + \beta_2x_2 + \beta_3x_1x_2 + \beta_4x_1^2 + \beta_5x_2^2 + \varepsilon$$

The first term is the intercept term, the second and third terms are the linear terms, the fourth term is the interaction term, and the fifth and sixth terms are the second degree terms other than interaction.

Suppose we have a complete and a reduced model and SSE_R comes from the reduced model and SSE_C and MSE_C come from the complete model. We may then test a set of betas equal to zero by using the following test statistic:

$$F = \frac{(\text{SSE}_R - \text{SSE}_C)/\# \text{ of } \beta\text{s tested in } H_0}{\text{MSE}_C}$$

Nonparametric Statistics

5-1 Distribution-free Tests

The tests discussed so far in this book assumed that the distributions from which we selected our samples were normally distributed. The tests with two or more samples also assumed that the populations had equal variances. However, there are situations in which our population distributions are clearly non-normal and where we do not have equal variabilities. There is a class of tests that test hypotheses about the location of populations and are free of the parameters of the distributions. These tests are called *distribution-free tests* or *nonparametric tests*. The hypotheses are stated in a different form than for parametric tests. They also do not have the normality assumptions. Rather than testing that population means are different, as was the case with two sample tests or with ANOVAs, we test that the location of the populations differs.

The null hypothesis states that the populations have the same distribution or the same location. This is illustrated in Fig. 5-1. The research hypothesis states that the population distributions differ or that population 1 is located to the right or left of population 2. These all correspond to one- or two-tailed tests. Figures 5-2 and 5-3 illustrate these situations; in these figures, population 1 is shown as the dashed curve and population 2 as the solid curve.

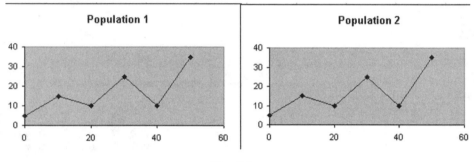

Fig. 5-1.

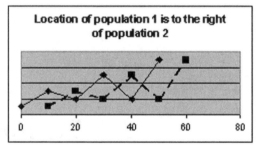

Fig. 5-2.

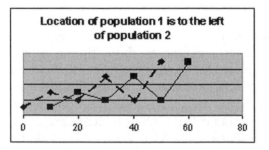

Fig. 5-3.

In nonparametric test procedures, the data is often replaced by signs or ranks or both signs and ranks. The signs and ranks are analyzed instead of the data itself. We often hear the complaint that nonparametric tests are wasteful of information because the original data is not used directly in reaching conclusions about populations, but rather signs and ranks are used in reaching conclusions.

5-2 The Sign Test

Purpose: To be used in situations where the assumptions of the paired t-test are not satisfied or where the data are ordinal.

Assumptions: The assumptions of the binomial distribution for X, the number of positive signs or negative signs in the sign test. When n is large ($np \geq 5$ and $nq \geq 5$), the test statistic $z = (x - np)/\sqrt{np(1 - p)}$ may be used. Otherwise, use the binomial distribution.

EXAMPLE 5-1
The sign test is used to analyze the results of a taste test. Suppose each of 20 individuals is asked to taste and state their preference for either brand A cola or brand B cola. The cola presented to the individual first is randomly determined and is presented to the individual to express his or her preference without knowing which cola is being tested at a given time. For example, John Doe is presented both colas in a plain container and he states his preference for cola A or cola B. This is done by all twenty of the participants. The null hypothesis is there is no difference in the two colas and the alternative is that a difference exists. Suppose cola A is chosen by 15 of the participants. The two-tailed p-value is $2P(x \geq 15)$. The determination of the p-value using Minitab is shown in Fig. 5-4.

SOLUTION
Assuming the null hypothesis to be true, $p = 0.5$. The pull-down **Calc** $\Rightarrow$ **Probability distribution** $\Rightarrow$ **Binomial** gives the binomial distribution dialog box, which is filled in as shown.

   Cumulative Distribution Function
Binomial with n = 20 and p = 0.500000

```
x            P(X <= x)
14.00        0.9793
```

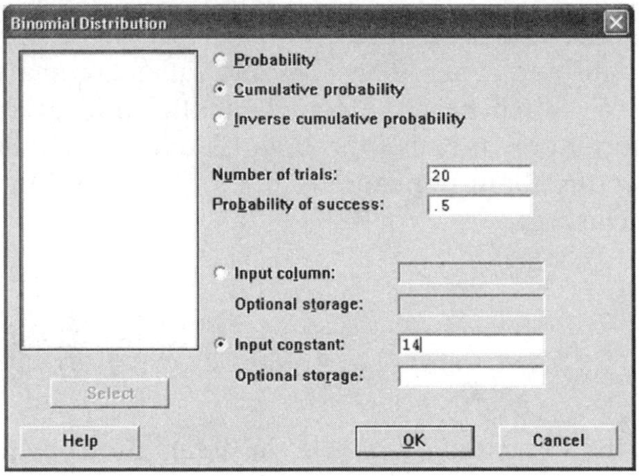

Fig. 5-4.

The probability $P(x \geq 15) = 1 - P(x \leq 14) = 1 - 0.9793 = 0.0207$ and the p-value $= 2P(x \geq 15) = 2(0.0207) = 0.0414$.

At $\alpha \leq 0.05$ the null would be rejected and the alternative hypothesis accepted. It would be concluded that cola A is preferred over cola B.

EXAMPLE 5-2

Use Excel to perform the test that there is no difference in preference.

SOLUTION

The determination using Excel is shown in Fig. 5-5. In order to get this dialog box, click Paste function and choose Statistical as function category; then, as function, choose Binomdist.

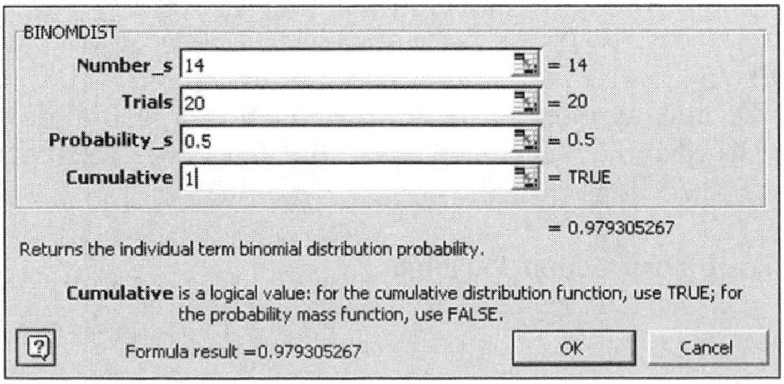

Fig. 5-5.

The function Binomdist is filled out as shown in Fig. 5-5. The cumulative probability $P(x \leq 14)$ is shown to be 0.979305267 in the dialog box. The probability $P(x \geq 15) = 1 - P(x \leq 14) = 1 - 0.9793 = 0.0207$ and the p-value $= 2P(x \geq 15) = 2(0.0207) = 0.0414$.

EXAMPLE 5-3

The sign test is used to test that the median of a population equals some value. A high school principal hypothesizes that the median time spent on the Internet per week by students at her school is greater than 10 hours. A sample of size 25 is taken and the weekly times spent on the Internet are: 3, 3, 4, 4, 4, 5, 5, 5, 5, 5, 10, 10, 13, 13, 13, 15, 15, 15, 17, 17, 25, 30, 30, 35, and 40 hours. The null hypothesis is H_0: median $= 10$ hours and the research hypothesis is H_a: median > 10 hours. Assuming the null hypothesis to be true, the original data is replaced by the following: $-, -, -, -, -, -, -, -, -, -, 0, 0, +, +, +, +, +, +, +, +, +, +, +, +, +$. There are $n = 23$ non-zero signs. Ten are negative and 13 are positive. Thirteen out of 23 are supportive of the research hypothesis. Assuming the null hypothesis to be true, the probability of 13 or more positive values out of 23 can be found using Minitab as follows.

SOLUTION

Cumulative Distribution Function
Binomial with n = 23 and p = 0.500000

```
x              P(X <= x)
12.00          0.6612
```

p-value $= P(X \geq 13) = 1 - 0.6612 = 0.3388$. Because the p-value is greater than 0.05, we are unable to reject the null hypothesis.

EXAMPLE 5-4

The sign test can also be used when your data is paired. Fifteen intersections are involved in an experiment. The number of accidents is recorded in a one-month period before and after the installation of a traffic control device at each intersection. The data is shown in the Table 5-1.

Table 5-1   Number of accidents per month before and after traffic control device installed.

Before	4	5	4	3	4	0	6	5	4	3	3	5	2	1	2
After	1	2	0	2	3	1	3	4	3	3	4	3	1	2	2
Sign	+	+	+	+	+	−	+	+	+	0	−	+	+	−	0

The null hypothesis is H_0: The number of accidents is the same before as after the installation of the traffic control device. The research hypothesis is H_a: The number of accidents is reduced by the installation of the traffic control device. There are 13 non-zero differences. If the null hypothesis is true, you would expect about 6.5 accidents before and 6.5 after. The more the split differs from this, the more you will doubt the null hypothesis. The data gives 10 plus signs and 3 negative signs. The p-value is equal to $P(X \geq 10)$.

SOLUTION
The calculation of the p-value using Excel proceeds as shown in Fig. 5-6.

Microsoft Excel - Book1				

File   Edit   View   Insert   Format   Tools   Data   Window   Help

A1   =   =BINOMDIST(9,13,0.5,1)

	A	B	C	D	E
1	0.953857				
2					

Fig. 5-6.

The p-value is equal to $P(X \geq 10) = 1 - P(X \leq 9) = 1 - 0.953857 = 0.046$. At $\alpha = 0.05$, we would reject the null hypothesis and conclude that the installation of the device reduced the number of accidents.

5-3 The Wilcoxon Rank Sum Test for Independent Samples

Purpose: To be used in situations where the independent samples t-test for two samples assumptions are not satisfied.

Assumptions: The two samples are random and independent. Also, the two population distributions from which the samples are drawn are continuous.

EXAMPLE 5-5
First we shall give an example to see the logic behind the Wilcoxon rank sum test. Then we will use statistical software to make these calculations for us.

Suppose we wish to test the following hypotheses at $\alpha = 0.05$, using two samples of size 3:

H_0: The two populations have the same location

H_a: The location of population 1 is to the right of
    the location of population 2

SOLUTION

Suppose our samples are as shown in Table 5-2. The two samples are combined and ranked together. The smallest number is assigned rank 1,

Table 5-2

Sample 1	Rank	Sample 2	Rank
30	5	20	1
33	6	24	2
26	3	28	4
	$T_1 = 14$		$T_2 = 7$

the next number rank 2, and so forth until all six numbers are ranked. The rank sum for sample 1 is T_1 and the rank sum for sample 2 is T_2. The data does support the research hypothesis. The strength of that support is given by the p-value, which will now be derived. Population 1 is yielding data that tends to indicate that population 1 is to the right of population 2. To see how significant this assignment of ranks is, we need to look at the set of all possible ranks. Remember, under the null hypothesis, all rank assignments are equally likely. Table 5-3 shows all possible rank assignments, with three ranks assigned to each sample.

Suppose we choose $T = T_1$ as our test statistic. The sampling distribution of T is given in Table 5-4 and is obtained from Table 5-3. Under the null hypothesis, each of the outcomes in Table 5-3 is equally likely.

The p-value is the probability of the sample we obtained, or one more supportive of the research hypothesis; that is, $p\text{-value} = P(T \geq 14) = P(T = 14$ or $T = 15) = 1/20 + 1/20 = 2/20 = 0.10$.

Consider the Minitab computation of the p-value. First the data is placed into columns C1 and C2 as shown in Fig. 5-7. The pull-down sequence

Table 5-3   All possible ranks with each sample size equal to three.

Ranks of sample 1	Rank sum	Ranks of sample 2	Rank sum	Ranks of sample 1	Rank sum	Ranks of sample 2	Rank sum
123	6	456	15	234	9	156	12
124	7	356	14	235	10	146	11
125	8	346	13	236	11	145	10
126	9	345	12	245	11	136	10
134	8	256	13	246	12	135	9
135	9	246	12	256	13	134	8
136	10	245	11	345	12	126	9
145	10	236	11	346	13	125	8
146	11	235	10	356	14	124	7
156	12	234	9	456	15	123	6

Table 5-4   Distribution of the Wilcoxon rank sum statistic for $n_1 = 3$ and $n_2 = 3$.

T	6	7	8	9	10	11	12	13	14	15
$P(T)$	1/20	1/20	2/20	3/20	3/20	3/20	3/20	2/20	1/20	1/20

Stat $\Rightarrow$ Nonparametric $\Rightarrow$ Mann–Whitney gives Fig. 5-8. This dialog box produces the following output.

Mann–Whitney Test and CI: sample1, sample2

```
sample1   N = 3   Median = 30.00
sample2   N = 3   Median = 24.00
Point estimate for ETA1-ETA2 is 6.00
91.9 Percent CI for ETA1-ETA2 is (-2.00, 13.00)
```

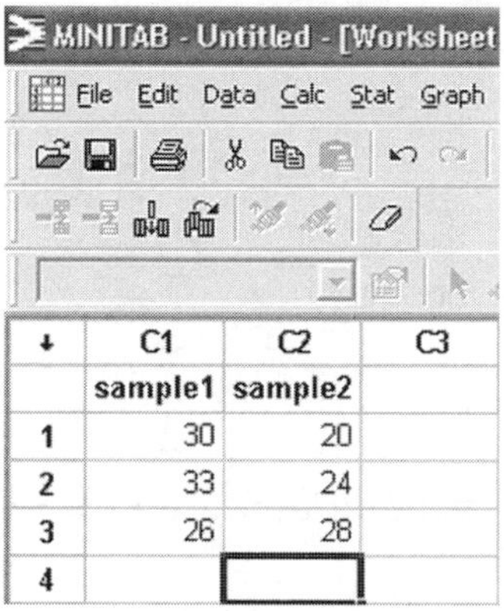

Fig. 5-7.

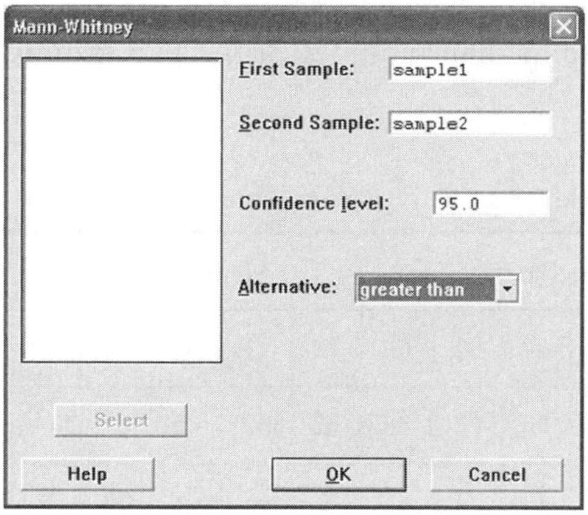

Fig. 5-8.

```
W = 14.0
Test of ETA1 = ETA2 vs ETA1 > ETA2 is significant at 0.0952
```

Cannot reject at alpha $= 0.05$

Note: The Mann–Whitney test is equivalent to the Wilcoxon rank sum test. Both tests are used when two independent samples are used and the normality assumption is in question so that the two-sample *t*-test is not appropriate.

When the two sample sizes are larger than 10, the T statistic is approximately normal with mean given by

$$\mu_T = \frac{n_1(n_1 + n_2 + 1)}{2}$$

and standard deviation given by

$$\sigma_T = \sqrt{\frac{n_1 n_2(n_1 + n_2 + 1)}{12}}$$

The approximation is not valid in the case $n_1 = 3$ and $n_2 = 3$.

EXAMPLE 5-6

A study compared a group of vegetarians with a group of meat-eaters over many years. One of the recorded data values was the lifetime of the individuals in the two groups. There were fifty in each group. The data is shown in Table 5-5. Use the Wilcoxon rank sum test to see whether there is a difference in the median lifetimes of the two groups.

Table 5-5   Lifetimes of vegetarians and meat-eaters.

Vegetarians					Meat-eaters				
80	67	92	69	78	71	85	78	63	66
65	86	88	88	72	80	35	87	74	79
89	89	74	25	92	63	78	77	78	67
70	74	88	69	77	67	72	65	81	65
87	76	90	78	45	69	75	90	72	76
61	69	87	55	75	62	70	55	82	70
94	69	95	71	70	67	71	69	56	63
98	61	66	84	91	62	78	91	58	63
81	72	71	90	73	71	80	30	65	76
70	77	83	71	67	70	45	71	75	70

SOLUTION

Mann–Whitney Test and CI: Vegetarians, Meat-eaters

```
Vegetari   N = 50          Median = 75.500
Meateate   N = 50          Median = 70.500
Point estimate for ETA1-ETA2 is    6.000
95.0 Percent CI for ETA1-ETA2 is (2.003,11.000)
W = 2918.0
Test of ETA1 = ETA2 vs ETA1 not = ETA2 is significant at 0.0068
The test is significant at 0.0068 (adjusted for ties)
```

The Mann–Whitney output indicates that the vegetarians do live longer than the meat-eaters at $\alpha = 0.05$. The sample difference is 5 years on the average. The *p*-value is 0.0068. The histograms in Figs. 5-9 and 5-10 indicate the non-normality of the data.

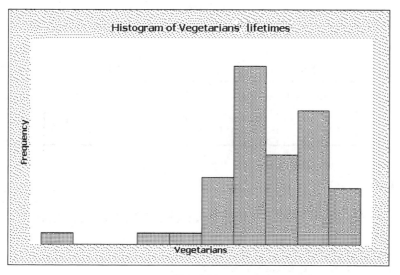

Fig. 5-9.

EXAMPLE 5-7

Two surgical techniques were compared with respect to the number of days the patients had to spend in the hospital after the surgeries. Table 5-6 shows the days required by the patients to be hospitalized following the surgeries.

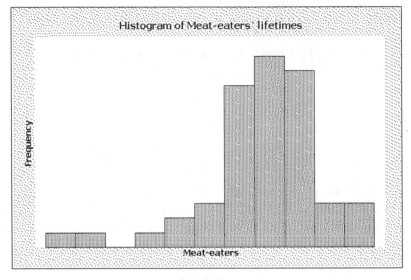

Fig. 5-10.

Table 5-6 Comparison of two surgical techniques in terms of hospital stays.

Technique 1	1	2	2	3	2	2	10	8	2	4
Technique 2	2	1	1	1	3	4	3	3	1	2

SOLUTION

Mann–Whitney Test and CI: Technique1, Technique2

```
Techniqu N = 10          Median = 2.000
Techniqu N = 10          Median = 2.000
Point estimate for ETA1-ETA2 is    1.000
95.5 Percent CI for ETA1-ETA2 is (-1.000,3.000)
W = 119.0
Test of ETA1 = ETA2 vs ETA1 not = ETA2 is significant at 0.3075
The test is significant at 0.2911 (adjusted for ties)
Cannot reject at alpha = 0.05
```

The study does not find a significant difference in the median times. However, technique 1 appears to result in long stays in about 20% of the cases. This might cause one to favor the second technique even though there is no difference between the two in terms of median hospital stay.

5-4 The Wilcoxon Signed Rank Test for the Paired Difference Experiment

Purpose: To be used when the normality assumption of the paired t-test is not satisfied.

Assumptions: The sample of differences is randomly selected from the population of differences. The probability distribution from which the sample of paired differences is drawn is continuous.

The data for this design are paired by taking pairs of measurements on the same experimental units, e.g. by using twins, doing a before/after study, or by pairing using husband/wives, etc. The assumption that the differences are normally distributed may not be satisfied and therefore the paired t-test procedure is not valid.

EXAMPLE 5-8

Suppose we wish to test the hypothesis that the time spent watching TV per week and the time spent on the Internet per week are different for junior high students. The data for ten students is shown in Table 5-7. The hypotheses to be tested are: H_0: Internet and TV times are the same; H_a: Internet and TV times differ. The test statistic may be either T^+ or T^-.

The absolute differences are ranked and the signs of the differences are retained. Both T^+ and T^- can be anything from 0 to 55.

> **Note:** The sum of T^+ and T^- is equal to the sum of the first $n = 10$ integers or $n(n+1)/2 = 10(11)/2 = 55$. If the null hypothesis is true, then both T^+ and T^- should be close to one-half of 55 or 27.5. The Minitab software will give the probability of obtaining a 19 and 36 split or something more extreme. That is just the p-value.

SOLUTION

The worksheet is filled out as shown in Fig. 5-11. The pull-down **Stat ⇒ Nonparametric ⇒ 1-sample Wilcoxon** gives the dialog box shown in Fig. 5-12. Fill it in as shown. The following output is produced.

Table 5-7   TV time and Internet time per week for 10 high school students.

Student	TV time	Internet time	D	$|D|$	Rank +	Rank −
Kidd	5	7	−2	2		1.5
Long	4	7	−3	3		3
Smith	10	2	8	8	8	
Jones	8	4	4	4	4.5	
Daly	9	3	6	6	6.5	
Lee	3	7	−4	4		4.5
Maloney	4	15	−11	11		9
Manley	5	7	−2	2		1.5
Liu	3	15	−12	12		10
Conley	5	11	−6	6		6.5
					$T^+ = 19$	$T^- = 36$

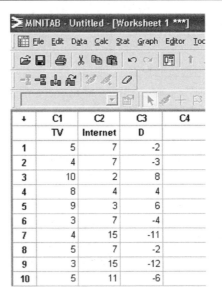

Fig. 5-11.

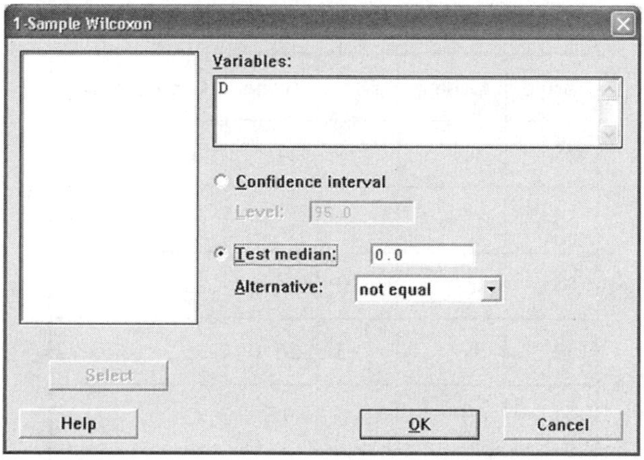

Fig. 5-12.

Wilcoxon Signed Rank Test: D

```
Test of median = 0.000000 versus median not = 0.000000

            N for    Wilcoxon                Estimated
     N      Test     Statistic      P        Median
D    10     10       19.0           0.415    -2.500
```

The p-value for the two-sided test is 0.415, indicating that the data does not indicate a difference in weekly TV time and weekly Internet time for junior high students.

If $n \geq 15$, then either T^+ or T^- may be assumed to have an approximate normal distribution with

$$\mu = \frac{n(n+1)}{4} \quad \text{and} \quad \sigma^2 = \frac{n(n+1)(2n+1)}{24}$$

EXAMPLE 5-9

A study recorded the ages of brides and grooms and the data is shown in Table 5-8. The null hypothesis is that of no difference in the two distributions and the research hypothesis is that there is a difference. The test is to be performed at $\alpha = 0.05$.

SOLUTION

The solution using Excel will be given first, followed by the Minitab solution.

The non-zero differences are entered into column A and the absolute values are calculated in column B by entering =ABS(A2) into B2 and then performing a click-and-drag operation in column B (Fig. 5-13). A sort is then performed on column B. Figure 5-14 shows the absolute values sorted.

Table 5-8 Bride and groom ages.

Bride	Groom	D	Bride	Groom	D
25	24	1	24	26	−2
30	35	−5	23	29	−6
35	35	0	24	25	−1
32	35	−3	26	35	−9
33	40	−7	30	32	−2
28	40	−12	34	33	1
27	22	5	33	35	−2
21	30	−9	35	35	0
22	23	−1	30	38	−8
34	35	−1	30	22	8

	A	B	C
1	D	ABS(D)	
2	1	1	
3	-5	5	
4	-3	3	
5	-7	7	
6	-12	12	
7	5	5	
8	-9	9	
9	-1	1	
10	-1	1	
11	-2	2	
12	-6	6	
13	-1	1	
14	-9	9	
15	-2	2	
16	1	1	
17	-2	2	
18	-8	8	
19	8	8	

Fig. 5-13.

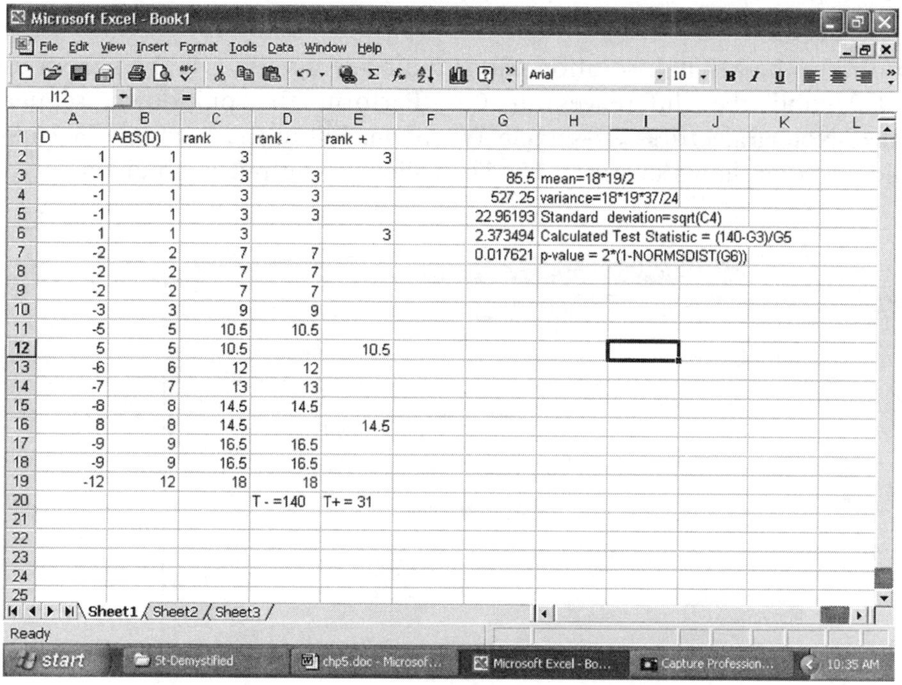

	A	B	C	D	E	G	H
1	D	ABS(D)	rank	rank -	rank +		
2	1	1	3		3		
3	-1	1	3	3		85.5	mean=18*19/2
4	-1	1	3	3		527.25	variance=18*19*37/24
5	-1	1	3	3		22.96193	Standard deviation=sqrt(C4)
6	1	1	3		3	2.373494	Calculated Test Statistic = (140-G3)/G5
7	-2	2	7	7		0.017621	p-value = 2*(1-NORMSDIST(G6))
8	-2	2	7	7			
9	-2	2	7	7			
10	-3	3	9	9			
11	-5	5	10.5	10.5			
12	5	5	10.5		10.5		
13	-6	6	12	12			
14	-7	7	13	13			
15	-8	8	14.5	14.5			
16	8	8	14.5		14.5		
17	-9	9	16.5	16.5			
18	-9	9	16.5	16.5			
19	-12	12	18	18			
20				T- =140	T+ = 31		

Fig. 5-14.

The ranks of the absolute differences are entered in column C. The difference 1 occurs five times. The ranks that would normally occur are 1, 2, 3, 4, and 5. The average of these five ranks is 3, which is therefore assigned to the first five 1s. Any time there is a tie for the ranks, the average of the ranks is assigned to the ties. The ranks for the negative differences are entered in column D and these for the positive differences are entered into column E. Then the ranks are summed to obtain T^+ and T^-.

> **Note:** In Fig. 5-14 the mean is calculated using $\mu = n(n+1)/4$, the variance is calculated using $\sigma^2 = n(n+1)(2n+1)/24$, the standard deviation is the square root of the variance, and the z value is calculated by $z = (T^+ - \mu)/\sigma$. The p-value is the area to the right of 2.37 doubled. Note that the p-value is = 2 * (1-NORMSDIST(G6)). The p-value is 0.018 when rounded to three places. We conclude that the brides are younger than the grooms.

EXAMPLE 5-10

Give the Minitab solution to Example 5-9.

SOLUTION
To obtain the Minitab solution, enter the bride ages in C1, the groom ages in C2, and the differences in C3. Perform the pull-down sequence **Stat ⇒ Nonparametrics ⇒ 1-sample Wilcoxon.** Fill in the 1-sample Wilcoxon dialog box as shown in Fig. 5-15. The following output is obtained.

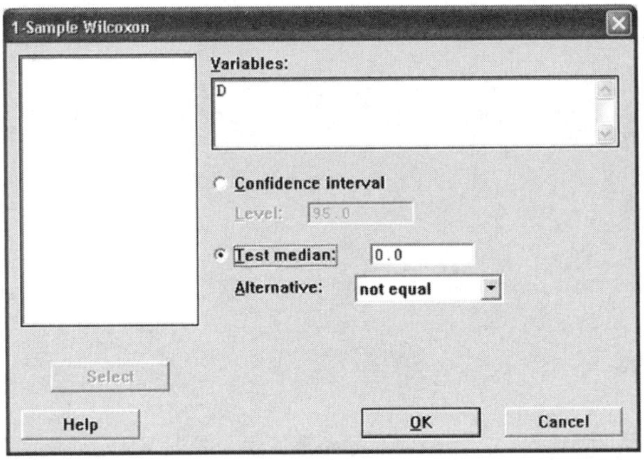

Fig. 5-15.

Wilcoxon Signed Rank Test: D
Test of median = 0.000000 versus median not = 0.000000

	N	N for Test	Wilcoxon Statistic	P	Estimated Median
D	20	18	140.0	0.019	2.500

Note that the Wilcoxon statistic is $T^- = 140$, the same that was obtained using Excel. The p-value is 0.019, compared to 0.018 in the Excel solution. The exact solution rather than the normal approximation may have been used in Minitab. This would have resulted in a slightly different p-value.

5-5 The Kruskal–Wallis Test for a Completely Randomized Test

Purpose: The Kruskal–Wallis test is an alternative to the completely randomized analysis of variance test procedure.

Assumptions: The samples are random and independent. There are five or more measurements in each sample. The probability distributions from which the samples are drawn are continuous.

When the means of k populations are compared and it is known that the populations do not have equal variances or that the populations are not normal, the Kruskal–Wallis nonparametric test is used. The data in the k samples are replaced by their ranks and the rank sums are analyzed.

EXAMPLE 5-11
Suppose we wished to compare three golf drivers. A golfer was asked to drive five golf balls with each of three different drivers. Fifteen golf balls of the same brand were randomly divided into three sets of five each. The golfer drove five with driver A, five with driver B, and five with driver C. The distances that the balls traveled in yards are shown in Table 5-9. Give the Excel solution to the problem.

Table 5-9   Distances driven by each of three drivers.

Driver A	Driver B	Driver C
250.8	253.2	245.8
254.2	255.4	265.5
252.3	254.4	255.7
255.6	255.0	266.7
253.5	254.0	270.4

SOLUTION
The Excel solution is shown in Fig. 5-16. The data is entered into column A and the treatment name (club) into column B. The data in column A is sorted from smallest to largest. The rank is then entered in column C. The rank sums are shown in F2, F3, and F5.

The test statistic is

$$H = \frac{12}{n(n+1)} \sum \frac{R_i^2}{n_i} - 3(n+1)$$

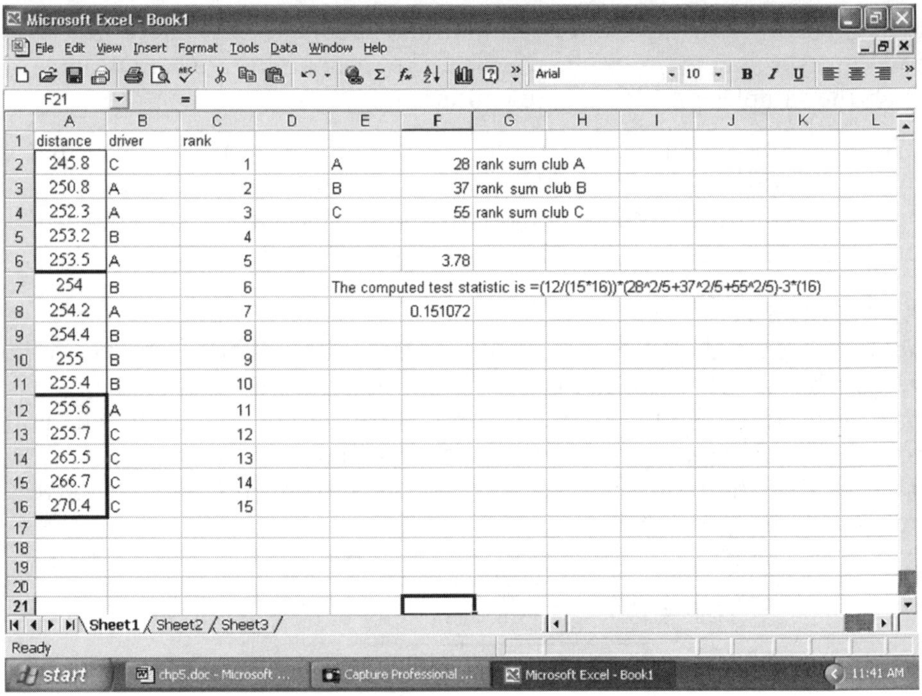

	A	B	C	D	E	F	G
1	distance	driver	rank				
2	245.8	C	1		A	28	rank sum club A
3	250.8	A	2		B	37	rank sum club B
4	252.3	A	3		C	55	rank sum club C
5	253.2	B	4				
6	253.5	A	5			3.78	
7	254	B	6		The computed test statistic is =(12/(15*16))*(28^2/5+37^2/5+55^2/5)-3*(16)		
8	254.2	A	7			0.151072	
9	254.4	B	8				
10	255	B	9				
11	255.4	B	10				
12	255.6	A	11				
13	255.7	C	12				
14	265.5	C	13				
15	266.7	C	14				
16	270.4	C	15				

Fig. 5-16.

where R_i = the rank sum for the ith sample, n_i = number of measurements in the ith sample, and n = total sample size. For sample sizes greater than or equal to 5, H has an approximate chi-square distribution with one less than the number of the population's degrees of freedom. The test statistic is evaluated in E6 as =(12/(15 * 16)) * (28 ^ 2/5 + 37^2/5 + 55^2/5) − 3 * (16). The p-value is evaluated in E7 as =CHIDIST(3.78,2). This gives the area under the curve to the right of 3.78. The 2 is the degrees of freedom. Since this is always an upper tailed test, this gives the p-value as =0.151. At $\alpha = 0.05$, there is no difference in the distances attainable with the three different drivers.

EXAMPLE 5-12

Give the Minitab solution to Example 5-11.

SOLUTION

The Minitab solution proceeds as follows. The data are entered into the worksheet as shown in Fig. 5-17.

The pull-down **Stat ⇒ Nonparametric ⇒ Kruskal-Wallis** gives the Kruskal–Wallis dialog box which is filled in as shown in Fig. 5-18. The output is as follows.

↓	C1	C2	C3
	Distance	Club	
1	250.8	1	
2	254.2	1	
3	252.3	1	
4	255.6	1	
5	253.5	1	
6	253.2	2	
7	255.4	2	
8	254.4	2	
9	255.0	2	
10	254.0	2	
11	245.8	3	
12	265.5	3	
13	255.7	3	
14	266.7	3	
15	270.4	3	

Fig. 5-17.

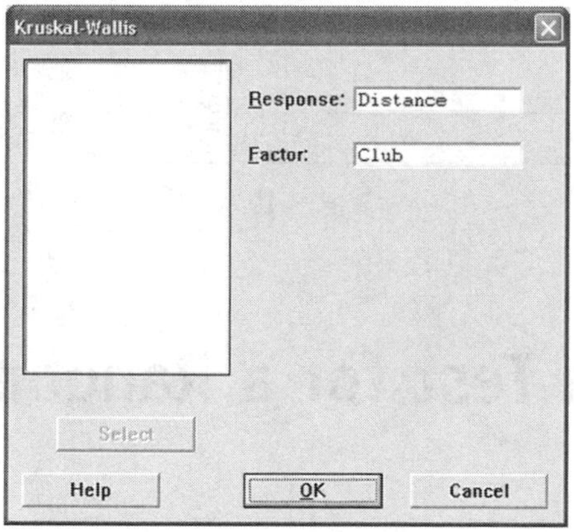

Fig. 5-18.

Kruskal-Wallis Test: Distance versus Club

Kruskal-Wallis Test on Distance

Club	N	Median	Ave Rank	Z
1	5	253.5	5.6	-1.47
2	5	254.4	7.4	-0.37
3	5	265.5	11.0	1.84
Overall	15		8.0	

$H = 3.78$ $DF = 2$ $P = 0.151$

The *p*-value indicates that the null should not be rejected at the usual alpha equal to 0.05. Figure 5-19 shows that the assumption of equal variances for each treatment, assumed when doing a parametric procedure, would be in doubt.

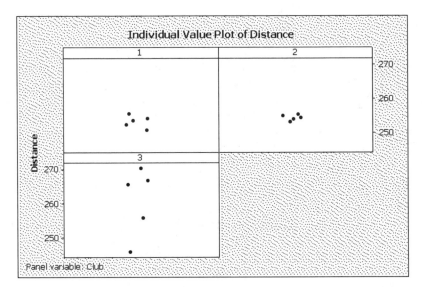

Fig. 5-19.

5-6 The Friedman Test for a Randomized Block Design

Purpose: The Friedman test provides a nonparametric alternative to analyzing a randomized block design.

Assumptions: The treatments are randomly assigned to experimental units within the blocks. The measurements can be ranked within the blocks. The probability distributions from which the samples within each block are drawn are continuous.

In Section 2-3 we introduced the parametric analysis of a block design. In this section we introduce the corresponding nonparametric analysis. Suppose we have *p* treatments to be applied in *b* blocks. The *p* treatments are randomly applied within each block and are replaced by their ranks within each block. The null and alternative hypotheses are

H_0: The locations of all p populations are the same

H_a: At least two population locations differ

The test statistic is

$$F_r = \frac{12}{bp(p+1)} \sum R_i^2 - 3b(p+1)$$

where $b =$ the number of blocks, $p =$ the number of treatments, and R_i is the rank sum for the ith treatment. The test statistic, F_r, is approximately chi-square with $p - 1$ degrees of freedom when either p or b is equal to or greater than 5. Like the Kruskal–Wallis test, the Friedman test is an upper tailed test. The test statistic tends to be large when the alternative hypothesis is true.

EXAMPLE 5-13

Four drugs (A, B, C, and D) are administered to five patients and enough time is allowed between administrations to allow for "wash out" effects. The reaction time is measured for each drug. Test the following hypotheses at $\alpha = 0.05$:

H_0: The populations of reaction times are identically distributed
for all four drugs

H_a: At least two of the drugs have different reaction times

The data are given in Table 5-10.

Table 5-10   Reaction times for five subjects for each of four drugs.

Subject	Drug A	Drug B	Drug C	Drug D
1	5.13	5.10	5.00	4.90
2	4.99	4.85	5.05	5.10
3	4.40	4.56	4.78	4.55
4	5.12	5.23	5.10	4.89
5	5.50	5.13	5.34	5.86

The Minitab solution will be given, followed by the Excel Solution.

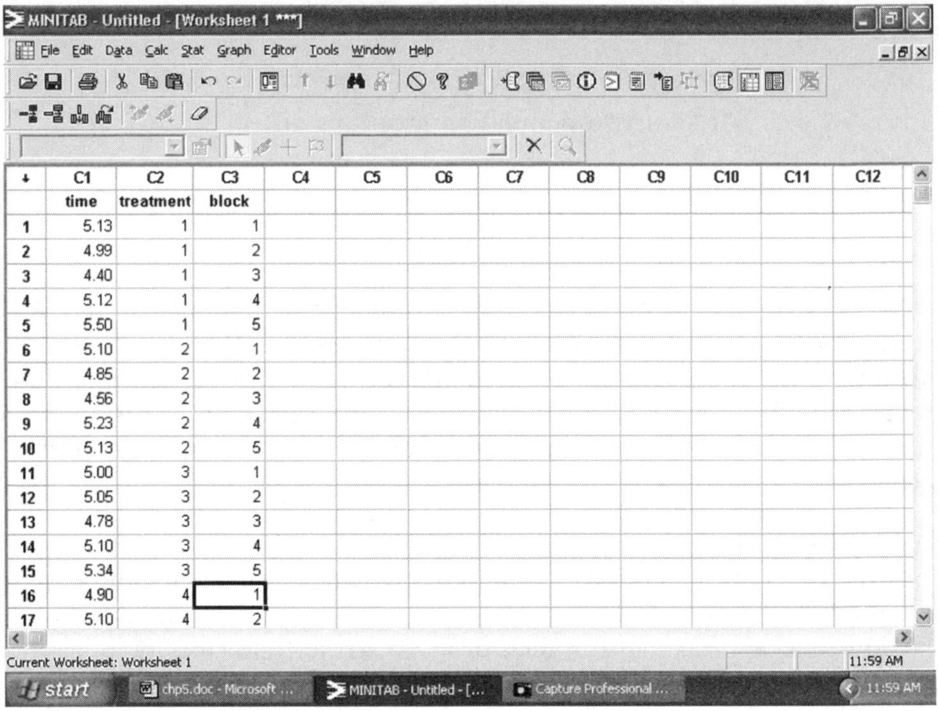

Fig. 5-20.

SOLUTION

Figure 5-20 shows how the data must be entered into the Minitab worksheet. Column C1 contains the reaction times, column C2 contains the treatment name, where A is coded 1, B is coded 2, C is coded 3, and D is coded 4, and column C3 contains the block or subject number. The pull-down **Stat ⟹ Nonparametric ⟹ Friedman** gives the dialog box shown in Fig. 5-21, which is filled out as shown. The following output is obtained.

Friedman Test: time versus treatment, block

```
Friedman test for time by treatment blocked by block
S = 0.12   DF = 3   P = 0.989
```

treatment	N	Est Median	Sum of Ranks
1	5	5.0225	13.0
2	5	4.9825	12.0
3	5	5.1075	13.0
4	5	5.0375	12.0

```
Grand median = 5.0375
```

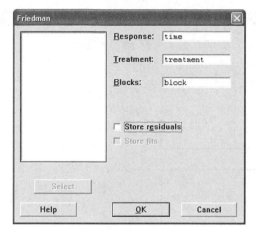

Fig. 5-21.

The closeness of the rank sums for the four drugs tells us there is no difference. Alternatively, the fact that the p-value $= 0.989$ is so large indicates that the null should not be rejected.

EXAMPLE 5-14
Give the Excel solution to Example 5-13.

SOLUTION
The Excel solution is shown in Fig. 5-22. The data are entered in A1:E6. The ranks are found in A9:E16. The test statistic is computed on the right-hand side of the worksheet. The expression $=$CHIDIST(0.12,3) gives the area to the right of 0.12 under the chi-square distribution having 3 degrees of freedom. This is the p-value.

Note that the Excel solution actually requires that the user shall have a complete understanding of how the test statistic is computed as well as the distribution of the test statistic. The Minitab solution actually carries out a lot of the work that the user must do him- or herself if using Excel.

EXAMPLE 5-15
In some cases the data are ordinal level and the ranks are given directly, as illustrated in the following example. Seven farmers were asked to rank the level of farm production constraint imposed by five conditions: drought, pest damage, weed interference, farming costs, and labor shortage. The rankings range from 1 (least severe) to 5 (most severe) and are shown in Table 5-11. Give the Minitab solution to this problem.

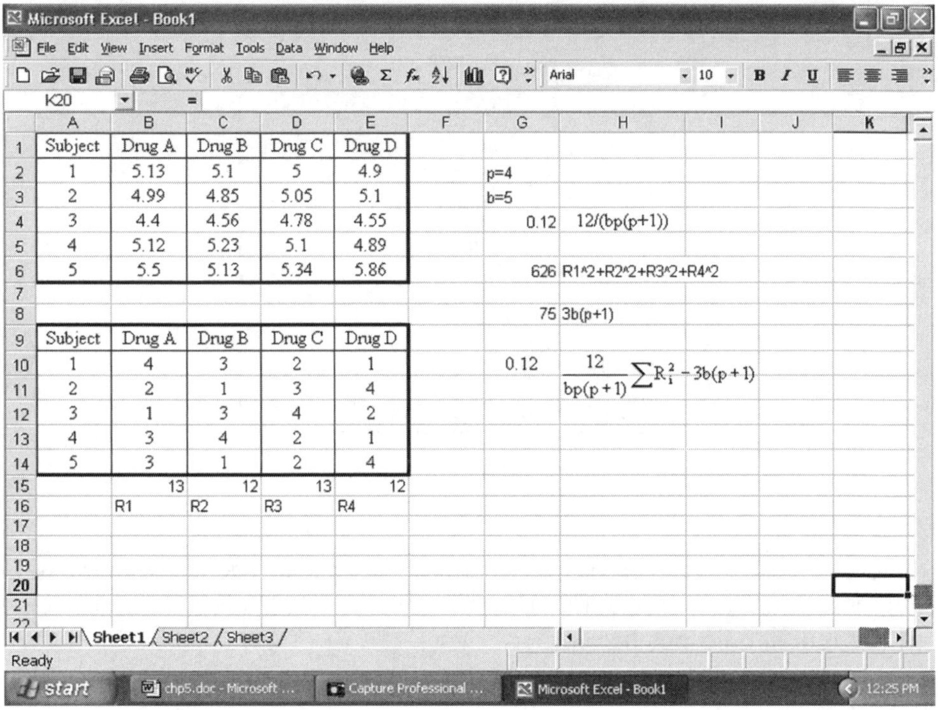

Subject	Drug A	Drug B	Drug C	Drug D
1	5.13	5.1	5	4.9
2	4.99	4.85	5.05	5.1
3	4.4	4.56	4.78	4.55
4	5.12	5.23	5.1	4.89
5	5.5	5.13	5.34	5.86

Subject	Drug A	Drug B	Drug C	Drug D
1	4	3	2	1
2	2	1	3	4
3	1	3	4	2
4	3	4	2	1
5	3	1	2	4
	13	12	13	12
	R1	R2	R3	R4

Fig. 5-22.

SOLUTION

The Minitab solution is as follows.

Friedman Test: Rank versus Trt blocked by Block

```
S = 16.57   DF = 4   P = 0.002

                                   Sum
                        Est        of
Trt        N            Median     Ranks
1          7            4.600      29.0
2          7            4.000      30.0
3          7            3.200      21.0
4          7            2.000      12.0
5          7            1.200      13.0

Grand median = 3.000
```

The p-value, 0.002, indicates that there is a difference between the five conditions. Drought and pest damage are the two chief production constraints.

Table 5-11   Seven farmers' rankings of farm production constraint caused by five conditions.

Farmer	Drought	Pest damage	Weed interference	Farming costs	Labor shortage
Smith	5	4	3	2	1
Jones	5	3	4	1	2
Long	3	5	4	2	1
Carter	5	4	1	2	3
Ford	4	5	3	2	1
Hartford	5	4	3	2	1
Farhatski	2	5	3	1	4
Rank sum	29	30	21	12	13

5-7 Spearman Rank Correlation Coefficient

Purpose: To provide a measure of the correlation between two sets of ranks.

Assumptions: The sample of experimental units on which the two variables are measured is randomly selected. The probability distributions of the two variables are continuous.

The sample Spearman rank correlation coefficient is represented by r_s and is found by ranking the two variables separately and then finding the parametric correlation between their ranks. The population measure is represented by ρ_s. When tied data occurs, the average of the ranks is assigned to those values. The Spearman measure of correlation is used when the bivariate data distribution is non-normal or when the data is ordinal level. The measurements are replaced by their ranks and then the parametric correlation is found between the ranks. This is the Spearman rank correlation coefficient.

EXAMPLE 5-16

Ten smokers were chosen and their average number of cigarettes smoked per day and their average diastolic blood pressures were determined. They are given in Table 5-12. Find the Spearman correlation coefficient between the two variables.

Table 5-12 Cigarettes smoked versus diastolic blood pressure.

Name	Number of cigarettes	Blood pressure
Livingstone	30	95
Kilpatrick	20	100
Jones	25	90
Smith	40	110
Ticer	20	95
Durham	25	93
Scroggins	35	85
Mosley	20	90
Daly	60	115
Langley	35	95

SOLUTION

First the data is replaced by its ranks. This is shown in Table 5-13. The ranks are then entered into columns C1 and C2 of the Minitab worksheet. The correlation of the ranks is obtained by the pull-down **Stat ⇒ Basic Statistics ⇒ Correlation**. The Spearman rank correlation coefficient is found, by using Minitab, to be as follows.

Correlations: C1, C2

```
Pearson correlation of C1 and C2 = 0.367
P-Value = 0.297
```

Table 5-13   Ranks of cigarettes smoked versus ranks of diastolic blood pressures.

Name	Number of cigarettes	Blood pressure
Livingstone	6.0	6.0
Kilpatrick	2.0	8.0
Jones	4.5	2.5
Smith	9.0	9.0
Ticer	2.0	6.0
Durham	4.5	4.0
Scroggins	7.5	1.0
Mosley	2.0	2.5
Daly	10.0	10.0
Langley	7.5	6.0

The null hypothesis is H_0: $\rho_s = 0$ versus H_a: $\rho_s > 0$. The p-value > 0.05 and we cannot conclude that a positive correlation exists between smoking and high blood pressure.

EXAMPLE 5-17
Use Excel to find the Spearman correlation coefficient for the data in Example 5-16.

SOLUTION
The Excel solution to the problem is shown in Fig. 5-23. The data is entered into columns A and B, with the number of cigarettes smoked per day in column A and the diastolic blood pressures in column B. The expression =RANK(A1,A$1:A$10,1) is entered into D1 and a click-and-drag is performed from D1 to D10. The expression =RANK(B1,B$1:B$10,1) is entered into E1 and a click-and-drag is performed from E1 to E10. The tied ranks are not assigned the average of the tied ranks by Excel and therefore need to be replaced. For example, 20 is the smallest value in column A. It occurs 3

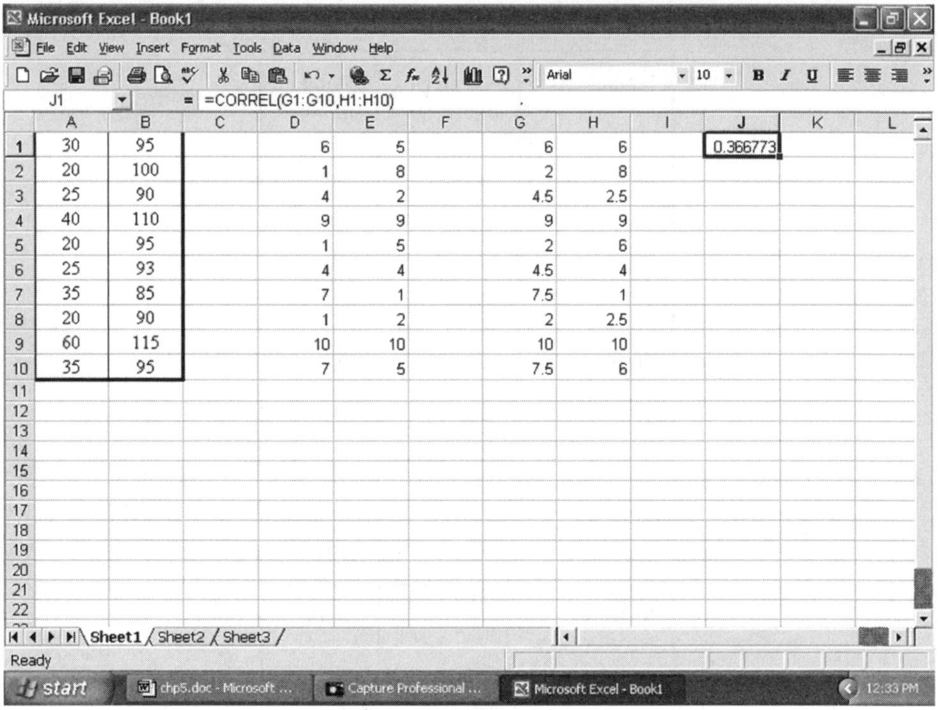

Fig. 5-23.

times and is assigned the rank 1. It should be assigned the average of ranks 1, 2, and 3 or 2. The adjustment is made and the rank replaced in column G. The two 4s in column D are replaced by 4.5 and the two 7s in D are replaced by 7.5 in column G. Similar adjustments are made in column E and replacements made in column H.

After the ranks are determined and put into G and H, the expression =CORREL(G1:G10,H1:H10) yields the correlation coefficient, with the ranks replacing the original data. That value is shown in cell J1. Tables may be consulted to determine whether the correlation is significant.

EXAMPLE 5-18
Ten teams of four programmers each competed in a programming contest. Two judges ranked the teams from 1 (best) to 10 (worst) according to their programs that gave solutions to a problem solved by all ten teams. The judges' rankings are given in Table 5-14. Test the following hypothesis.

$$H_0: \rho_s = 0 \text{ versus } H_a: \rho_s > 0$$

Table 5-14 Rankings of teams by judges 1 and 2.

Team	Judge 1	Judge 2
1	8	8
2	4	6
3	3	4
4	7	5
5	5	3
6	9	10
7	10	9
8	1	2
9	2	1
10	6	7

where ρ_s is the population correlation coefficient between the rankings of the two judges.

SOLUTION
The correlation is found, by using Minitab, to be the following.

Correlations: Judge 1, Judge 2
Pearson correlation of Judge 1 and Judge 2 = 0.891
P-Value = 0.001

We conclude that the judges' rankings are positively correlated (a good thing!).

Since the data in this example is given directly as ranks, the ranks are entered directly into an Excel spreadsheet and the CORREL function is used to find the correlation coefficient, as shown in Fig. 5-24.

Fig. 5-24.

Note: The significance of r_s may be determined by noting that $z = r_s\sqrt{n-1}$ has an approximately standard normal distribution. This approximate relationship will come in handy if you are using Excel to compute r_s and need to determine the significance of the outcome. The approximation becomes better as n becomes larger.

Suppose, for example, you find $r_s = 0.367$ as in Fig. 5-22. Then $z = r_s\sqrt{n-1} = 0.367\sqrt{10-1} = 0.367(3) = 1.10$ and the approximate p-value is $P(Z > 1.10) = 0.135$. This would tell you that the value of r_s is not significant at $\alpha = 0.05$.

5-8 Exercises for Chapter 5

1. Table 5-15 shows the miles per gallon obtained with 30 full tanks of gasoline. Use the sign test to test the null hypothesis that the median is 30.5. The alternative is that the median is not 30.5. Test at $\alpha = 0.05$.
2. A poll of 500 individuals was taken and they were asked which of two candidates they would vote for as governor in the coming election. The poll results were that 275 favored candidate A and 225 favored candidate B. Note that, before the poll was taken, neither one was favored. At $\alpha = 0.05$, can a winner be predicted?

Table 5-15 Miles per gallon for 30 compact automobiles.

26.7	32.6	27.7	29.4	30.0	28.0	28.2	24.0	31.2	28.2
25.4	28.7	24.7	28.6	25.8	28.3	29.3	26.3	30.3	28.4
29.1	33.1	27.5	28.8	32.5	27.9	33.4	28.7	22.5	26.9

3. Two methods of teaching algebra were compared. One method integrated Excel for performing certain algebraic operations and the other method taught algebra in the traditional way. The scores made on a common final exam given in both courses are given in Table 5-16. Use the Wilcoxon rank sum test to determine whether there is a difference in the two groups' scores. Test at $\alpha = 0.05$. Give the sum of ranks for both groups.

Table 5-16 Comparison of two methods of teaching algebra.

Experimental group					Traditional group				
69	72	75	67	71	74	79	70	75	76
67	75	69	68	67	84	72	65	71	77
70	72	65	71	71	80	80	71	76	77
60	72	75	67	64	69	76	75	72	79
70	65	68	58	80	78	76	70	79	83

4. The time spent per week watching TV was measured for 30 married couples. The data is given in Table 5-17. At $\alpha = 0.05$, test that husbands and wives watch equal amounts of TV per week. Use the Wilcoxon signed rank test. Give the values for T^- and T^+ and the p-value.

5. A study compared the four diets: low-carb high-protein Atkins diet, lean-meat Zone diet, Weight Watchers plan, and low-fat vegetarian Dean Ornish diet. Eighty overweight people were divided into four

Table 5-17   Comparison of TV watching time per week for 30 husbands and wives.

Husband	Wife	Husband	Wife	Husband	Wife
7	7	13	7	7	5
5	8	8	10	8	7
7	8	14	10	11	13
7	2	7	7	6	8
8	8	10	5	8	10
11	6	9	7	8	7
7	11	9	8	8	6
9	10	7	8	9	10
12	9	7	7	10	8
5	7	10	9	10	6

groups. The weight losses, one year after starting the diets, are given in Table 5-18. Use the Kruskal–Wallis procedure to test that the weight losses are the same for the four diets at $\alpha = 0.05$.

6. A study was designed to investigate the effect of animals on human stress levels. Five patients were used in the experiment. One time the finger temperature was taken with a dog in a room with the patient, one time the finger temperature was taken with a picture of a dog in the room, and a third time the finger temperature was taken with neither a dog nor a picture of a dog. Increasing finger temperature indicates an increased level of relaxation. Using the data in Table 5-19, test for differences in finger temperature due to the presence of a dog at $\alpha = 0.05$.

7. A panel of nutritionists and a group of housewives ranked ten breakfast foods on their palatability (Table 5-20). Calculate the Spearman rank correlation coefficient and test H_0: $\rho_s = 0$ versus H_a: $\rho_s > 0$ at $\alpha = 0.05$.

Table 5-18 Weight loss for each of the four diets.

Atkins	Zone	Wt. Watcher	Ornish
10	5	10	11
9	9	9	12
19	12	12	10
4	18	8	13
10	9	3	16
12	8	9	9
10	5	9	23
10	5	11	10
12	10	7	9
12	21	7	8
11	6	7	10
12	11	20	4
10	10	14	8
8	12	26	8
12	5	7	11
8	10	5	11
23	10	13	8
4	7	9	9
7	6	14	17
9	9	11	6

Table 5-19   Patients' finger temperatures with live dog, dog's photo, and neither.

Patient	Live dog	Dog photo	Neither
1	96.5	96.4	95.5
2	94.5	93.6	94.1
3	95.5	94.5	94.0
4	93.8	93.6	93.5
5	97.5	96.7	95.7

Table 5-20   Rankings of breakfast foods by nutritionists and housewives.

Breakfast food	Nutritionists	Housewives
Bagels	1	2
Eggs	3	1
Donuts	10	3
Bacon	5	4
Sausage	6	5
Ham	4	6
Whole wheat toast	2	7
Waffle	9	10
Pancake	8	9
Oats	7	8

5-9 Chapter 5 Summary

The *sign test* is used when the data consists of signs, either positive or negative. The number of positive (or negative) signs follows a binomial distribution. The binomial distribution of Excel or Minitab may be used to compute the p-value for a given test.

When two independent samples are selected and the normality assumptions are in doubt, the *Mann–Whitney Test*, which is equivalent to the *Wilcoxon rank sum test*, is used to analyze the data. The Minitab pull-down **Stat ⇒ Nonparametric ⇒ Mann-Whitney** is used to analyze two independent samples and is the nonparametric equivalent of the independent samples t-test.

The *Wilcoxon signed rank test* is used when the data are paired and at a level greater than ordinal. When the differences are at the interval or ratio level but not normally distributed, this is the test to use. The pull-down **Stat ⇒ Nonparametrics ⇒ 1-sample Wilcoxon** is applied to the differences. This test takes the magnitude of the differences into account when the sign test does not.

The *Kruskal–Wallis test* is used when the assumptions of a one-way analysis of variance are not satisfied. The Minitab pull-down **Stat ⇒ Nonparametric ⇒ Kruskal-Wallis** is used to perform a Kruskal–Wallis test procedure. An Excel procedure for doing the Kruskal–Wallis analysis is also given.

The *Friedman test for a randomized block design* is used when the assumptions for a randomized block design are not satisfied. The Minitab pull-down **Stat ⇒ Nonparametric ⇒ Friedman** is used to perform a Friedman test procedure. An Excel procedure for doing the Friedman analysis is also given.

The *Spearman correlation coefficient* calculates the correlation of the ranks. The original data is replaced by its ranks and the Pearson parametric measure of correlation is computed using the ranks instead of the original data. The Minitab pull-down **Stat ⇒ Basic Statistics ⇒ Correlation** is applied to the ranks of the original data.

CHAPTER

Chi-Squared Tests

6-1 Categorical Data and the Multinomial Experiment

A *binomial experiment* consists of a sequence of n trials. On each trial, one of two outcomes can occur. The two outcomes are usually referred to as failure and success. If p_1 is the probability of success and p_2 is the probability of failure, then $p_1 + p_2 = 1$ and p_1 and p_2 do not change from trial to trial.

EXAMPLE 6-1

Suppose we flip a fair coin ten times and we are interested in the number of heads and tails that occur. In this case $p_1 = p_2 = 0.5$. We expect $np_1 = 10(0.5) = 5$ heads and $np_2 = 10(0.5) = 5$ tails to occur. Now suppose we perform the experiment and n_1 heads and n_2 tails occur. The test statistic $(n_1 - np_1)^2/np_1 + (n_2 - np_2)^2/np_2$ has an approximate chi-squared distribution with 1 degree of freedom when $np_1 \geq 5$ and $np_2 \geq 5$. Suppose we wish to test the following hypotheses concerning the coin:

H_0: $p_1 = 0.5$, $p_2 = 0.5$ $\qquad$ H_a: The probabilities are not equal to 0.5

SOLUTION
You flip the coin and obtain 9 heads and 1 tail. The test statistic equals the following (assuming the null hypothesis is true): $(9 - 5)^2/5 + (1 - 5)^2/5) = 6.4$. The p-value is the area to the right of 6.4 on a chi-squared curve having 1 degree of freedom. Using Minitab, we find the following.

```
Chi-Square with 1 DF
x            P(X <= x)
6.4          0.988588
```

The p-value is $1 - 0.988588 = 0.011412$. We would reject that the coin is fair at $\alpha = 0.05$, since the p-value is less than 0.05.

A *multinomial experiment* has K outcomes possible on each trial rather than only two as in the case of the binomial. The following are the properties of a multinomial experiment.

1. The experiment has n identical trials.
2. There are K possible outcomes to each trial. These outcomes are referred to as classes, categories, or cells.
3. The probabilities of the K outcomes are represented by $p_1, p_2, \ldots, p_K$ and they remain the same from trial to trial. Furthermore, $\sum_{i=1}^{K} p_i = 1$.
4. The trials are independent.
5. The cell counts $n_1, n_2, \ldots, n_K$ are random variables. They represent the number of observations that fall into each of the K categories.

The binomial experiment is a special case of a multinomial experiment with $K = 2$ possible outcomes to each trial. We shall give a few examples of multinomial experiments.

EXAMPLE 6-2
Suppose a die is rolled 120 times and we count the number of times 1, 2, 3, 4, 5, and 6 occur in the 120 rolls. This is an example of a multinomial experiment.

SOLUTION
The experiment consists of 120 trials (rolls). There are six possible outcomes (1, 2, 3, 4, 5, or 6) to each trial. The probabilities of the six outcomes are denoted by p_1, p_2, p_3, p_4, p_5, and p_6 and remain the same from trial to trial, and the sum of the six p values must equal 1. The rolls are independent of one another. The cell counts n_1, n_2, n_3, n_4, n_5, and n_6 are the random variables of interest and have a multinomial distribution.

EXAMPLE 6-3
Suppose that, among Internet users, 30% send 5 or fewer e-mails per week, 40% send more than 5 but less than 20 e-mails per week, and 30% send 20

or more per week. Suppose we randomly survey 250 Internet users and determine the number of e-mails per week that each sends.

SOLUTION
This may be viewed as a multinomial experiment. The experiment consists of 250 trials (surveys). There are three possible outcomes (≤ 5, $5 <$ number of e-mails < 20, ≥ 20). The probabilities ($p_1 = 0.3$, $p_2 = 0.4$, $p_3 = 0.3$) remain the same from trial to trial and the sum of the probabilities equals 1. Since we randomly select the Internet users, their responses are independent of each other. The cell counts n_1, n_2, and n_3 are the random variables of interest, and have a multinomial distribution.

6-2 Chi-Squared Goodness-of-Fit Test

Purpose: To test a hypothesis about a set of categorical probabilities for a multinomial variable.

Assumptions: A multinomial experiment has been conducted. We ensure this is the case by taking a random sample from the population. Every cell has an expected count of five or more.

EXAMPLE 6-4
The hypothesis of interest in a goodness-of-fit test concerns the probabilities associated with the K outcomes. Consider the experiment of rolling a die 120 times and counting the number of times that each of the six faces turns up. The null hypothesis is that the die is fair and the research hypothesis is that the die is unfair. The hypotheses are

$$H_0: p_1 = p_2 = p_3 = p_4 = p_5 = p_6 = 1/6$$
$$H_a: \text{At least one } p_i \text{ is not equal to } 1/6$$

Find the solution using Minitab.

SOLUTION
If the null hypothesis is true, the expected frequency will equal $e_i = np_i = 120(1/6) = 20$ for each cell ($i = 1, 2, 3, 4, 5, 6$). Let f_i ($i = 1, 2, 3, 4, 5, 6$) represent the observed frequencies. If each expected frequency is 5 or greater, then

$$\chi^2 = \sum_{i=1}^{6} \frac{(f_i - e_i)^2}{e_i}$$

has an approximate chi-squared distribution with $K - 1$ degrees of freedom. If the null hypothesis is true, then the f_i will be near the e_i and the test statistic will be close to 0. Otherwise, the test statistic will tend to be large. Therefore the null is rejected only for large values of the test statistic. Suppose that 10 ones, 25 twos, 30 threes, 20 fours, 30 fives, and 5 sixes are observed.

The Minitab solution proceeds as follows. The expected numbers are entered into C1 and the observed numbers are entered into C2 as shown in Fig. 6-1. The pull-down **Calc** $\Rightarrow$ **Calculator** gives the Calculator dialog box, which is filled in as shown in Fig. 6-2. This gives the results shown in Fig. 6-3.

↓	C1	C2	C3
	expected	observed	
1	20	10	
2	20	25	
3	20	30	
4	20	20	
5	20	30	
6	20	5	
7			

Fig. 6-1.

The pull-down **Calc** $\Rightarrow$ **Column Statistics** gives Fig. 6-4 and clicking OK gives the sum.

Sum of C3
Sum of C3 = 27.5

This value, 27.5, is the test statistic. The pull-down **Calc** $\Rightarrow$ **Probability Distribution** $\Rightarrow$ **Chi-Square** gives the following.

Cumulative Distribution Function
Chi-Square with 5 DF

```
x            P(X <= x)
27.5         0.999954
```

The p-value is $1 - 0.999954 = 0.000046$.

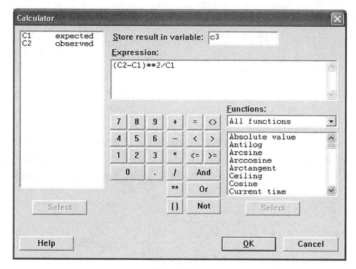

Fig. 6-2.

↓	C1	C2	C3	C4
	expected	observed		
1	20	10	5.00	
2	20	25	1.25	
3	20	30	5.00	
4	20	20	0.00	
5	20	30	5.00	
6	20	5	11.25	
7				

Fig. 6-3.

EXAMPLE 6-5
Find the solution to Exercise 6-4 using Excel.

SOLUTION
The Excel solution is given in Figs. 6-5 and 6-6. **CHITEST** is chosen from the Paste function dialog box, as shown in Fig. 6-5. The expected values are entered into column A and the observed values are entered into column B. The p-value is shown in Fig. 6-6 and equals $4.55759E - 05$ or 0.0000455759. The Minitab or Excel solution results in a rejection of the null hypothesis and a conclusion that the die is not balanced.

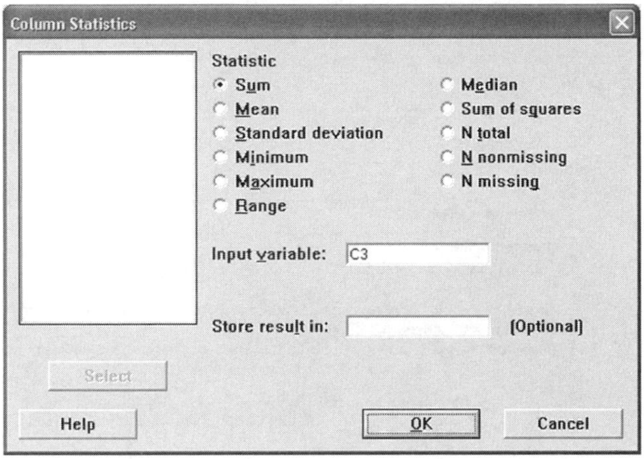

Fig. 6-4.

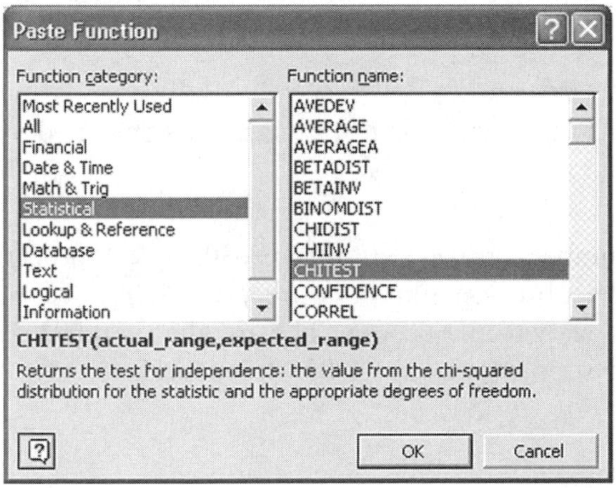

Fig. 6-5.

EXAMPLE 6-6

Suppose a survey reports that, among business Internet users, 30% send 5 or fewer e-mails per week, 40% send more than 5 but less than 20 e-mails per week, and 30% send 20 or more per week. In order to see whether the percents hold at Ace Technology, a survey of 100 employees is conducted. The results are: 25 send 5 or fewer, 35 send between 5 and 20, and 40 send 20 or more. The null hypothesis is that Ace Technology sends the same

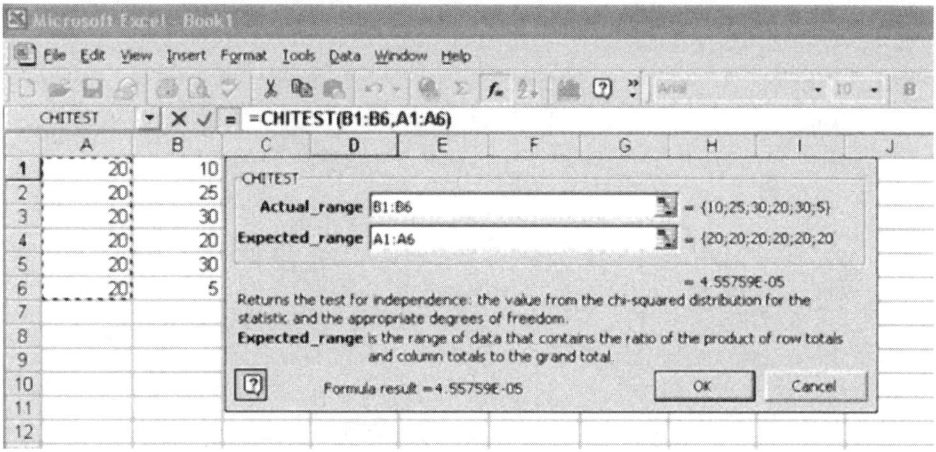

Fig. 6-6.

proportion as is reported by the business Internet users and the research hypothesis is that the distribution is different. The hypotheses are:

$$H_0: p_1 = 0.3, \quad p_2 = 0.4, p_3 = 0.3 \quad \text{and}$$

$$H_a: \text{At least one } p_i \text{ is not equal to its specified value}$$

SOLUTION

The expected numbers are $e_1 = 100(0.3) = 30$, $e_2 = 100(0.4) = 40$, and $e_3 = 100(0.3) = 30$. The observed numbers are $f_1 = 25$, $f_2 = 35$, and $f_3 = 40$. The Excel solution is shown in Fig. 6-7. The p-value is 0.091 and the research hypothesis is not supported at the 0.05 alpha level.

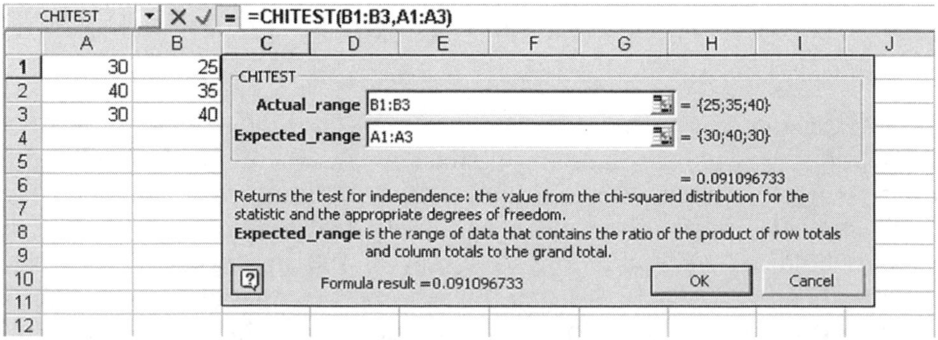

Fig. 6-7.

6-3 Chi-Squared Test of a Contingency Table

Purpose: To determine whether two categorical variables are independent or not.

Assumptions: The n observed counts are a random sample from the population being investigated. This may be considered a multinomial experiment with $r \times c$ possible outcomes. The sampling must be large enough so that the expected count, e_{ij}, is 5 or more for every value of i and j.

EXAMPLE 6-7
The chi-squared test of a contingency table allows us to test the independence of two variables, both of which may be at the nominal level. Suppose one variable is political party affiliation: Democrat, Republican, and Independent, and the other variable is tax policy: cut taxes or raise taxes. The null hypothesis is H_0: The two variables are independent, and the research hypothesis is H_a: The two variables are dependent. Suppose the results of a survey are as shown in the Table 6-1.

Table 6-1   Contingency table of political affiliation versus tax policy.

	Political affiliation		
Tax policy	**Democrat**	**Republican**	**Independent**
Cut taxes	25	55	20
Raise taxes	60	30	10

The test statistic for testing the above null hypothesis is

$$\chi^2 = \sum \frac{(f_{ij} - e_{ij})^2}{e_{ij}}$$

The summation is over all six cells of the above table. The f_{ij} are the observed values and the e_{ij} are the expected values if the null hypothesis is true and the two categorical variables are independent of one another.

If the expected value for each cell is 5 or greater, the test statistic has an approximate chi-squared distribution with $(r-1)(c-1)$ degrees of freedom. The number of rows is r and the number of columns is c in the contingency table. The hypothesis of independence is rejected for large values of the test statistic (that is, this is an upper tailed test only).

SOLUTION

The observed data are entered in the Minitab worksheet as shown in Fig. 6-8.

↓	C1	C2	C3	C4
	Democrats	Republicans	Independents	
1	25	55	20	
2	60	30	10	
3				
4				

Fig. 6-8.

The pull-down **Stat** ⇒ **Tables** ⇒ **Chi-Square Test** gives the dialog box shown in Fig. 6-9.

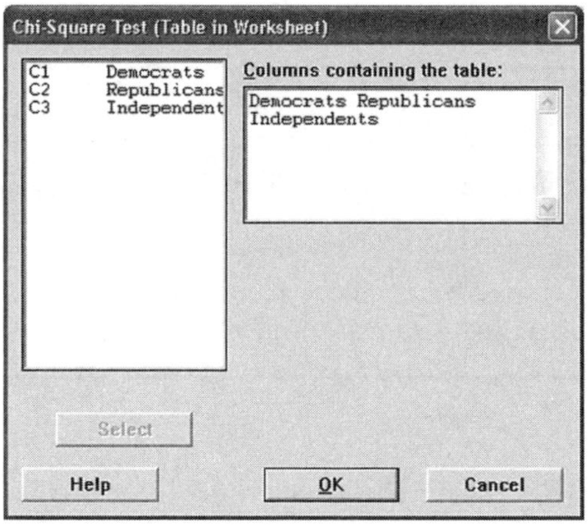

Chi-Square Test (Table in Worksheet)

C1	Democrats
C2	Republicans
C3	Independent

Columns containing the table:

Democrats Republicans Independents

Select

Help   OK   Cancel

Fig. 6-9.

The output is:

Chi-Square Test: Democrats, Republicans, Independents

```
Expected counts are printed below observed counts
Chi-Square contributions are printed below expected counts
```

	Democrats	Republicans	Independents	Total
1	25	55	20	100
	42.50	42.50	15.00	
	7.206	3.676	1.667	
2	60	30	10	100
	42.50	42.50	15.00	
	7.206	3.676	1.667	
Total	85	85	30	200

Chi-Sq = 25.098, DF = 2, P-Value = 0.000

The *p*-value tells us that political affiliation and tax cut opinion are related; that is, the null hypothesis of independence is rejected.

Note: The expected frequency of the cell in column *j* and row *i* is

$$e_{ij} = \frac{\text{column } j \text{ total} \times \text{row } i \text{ total}}{\text{sample size}}$$

EXAMPLE 6-8
Solve the above example using Excel.

SOLUTION
The Excel solution proceeds as follows. The observed values are copied into column A with the columns copied on top of each other. In column B the corresponding expected values are copied. Then, in C1, the expression =(A1-B1)^2/B1 is entered and a click-and-drag is performed from C1 to C6. In C7, the expression =SUM(C1:C6) performs the summation over all cells. The *p*-value is the area to the right of 25.09804 under the chi-squared curve having $(r - 1)(c - 1) = (2 - 1)(3 - 1) = 2$ degrees of freedom. This is shown in the dialog box (Fig. 6-10). The *p*-value is 0.00000354838.

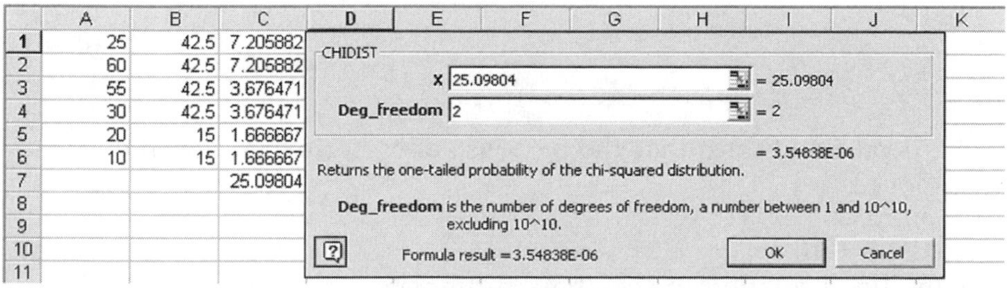

Fig. 6-10.

EXAMPLE 6-9

Suppose a large company has solicited its employees about its pension plan. The breakdown of responses from the employees is shown in Table 6-2. Test to see whether employee category is independent of opinion concerning the pension plan at $\alpha = 0.05$.

Table 6-2   Employee responses for three groups of employees.

Responses	Category of employee		
	Blue-collar	**White-collar**	**Manager**
For	75	35	12
Against	25	15	8

SOLUTION

The Minitab solution is given first.

Chi-Square Test: blue-collar, white-collar, manager

Expected counts are printed below observed counts
Chi-Square contributions are printed below expected counts

```
           blue-collar    white-collar    manager         Total
1          75             35              12              122
           71.76          35.88           14.35
            0.146          0.022           0.386

2          25             15              8               48
           28.24          14.12           5.65
            0.371          0.055           0.980

Total      100            50              20              170
```

Chi-Sq = 1.960, DF = 2, P-Value = 0.375

The p-value indicates that the response does not depend on the category of employee.

EXAMPLE 6-10

Give the Excel solution to the above example.

SOLUTION

The Excel solution is shown in Fig. 6-11. Again, the *p*-value indicates that the response does not depend on the category of employee.

	A	B	C	D	E	F	G	H	I	J	K
1	75	71.76	0.146288	CHIDIST							
2	25	28.24	0.371728		X	1.95672			= 1.95672		
3	35	35.88	0.021583								
4	15	14.12	0.054844	Deg_freedom	2				= 2		
5	12	14.35	0.384843								
6	8	5.65	0.977434						= 0.375927122		
7			1.95672	Returns the one-tailed probability of the chi-squared distribution.							
8											
9				Deg_freedom is the number of degrees of freedom, a number between 1 and 10^10, excluding 10^10.							
10				?	Formula result =0.375927122			OK		Cancel	
11											

Fig. 6-11.

6-4 Exercises for Chapter 6

1. Historically, grades in Dr Stephens' statistics class have followed the distribution: 10% As, 20% Bs, 50% Cs, 15% Ds, and 5% Fs. The number of grades in his various statistics classes the past year were: A = 20, B = 25, C = 60, D = 35, and F = 10. Test the following hypothesis: H_0: The current distribution is the same as in years past, versus H_a: The distribution differs from that of years past.

2. A die is rolled 90 times. On 35 times a *1 or 2* occurs, 40 times a *3 or 4* occurs, and the rest of the time a *5 or 6* occurs. Test the null hypothesis that the die is balanced versus the alternative that it is not balanced, at $\alpha = 0.05$.

3. A study looked at the relationship between marital status and religion. Test whether religious affiliation is independent of marital status. The data are given in Table 6-3. Give the test statistic, the *p*-value, and your conclusion for $\alpha = 0.05$.

Table 6-3   Contingency table of marital status versus religion.

Marital status	**Catholic**	**Protestant**	**Other**
Married	135	130	60
Divorced	40	50	25
Single	25	20	15

4. A quality control expert wants to determine whether there is dependence between the shift that manufactures compact disks and the quality of the disk. She gathers the following sample data (Table 6-4). Give the test statistic, the p-value, and your conclusion for $\alpha = 0.05$.

Table 6-4   Contingency table of CD quality versus shift.

	Shift		
CD quality	**1**	**2**	**3**
Good	185	175	170
Acceptable	55	60	65
Unacceptable	15	15	15
Rework	10	15	15

5. Social scientists say that interaction with family members steers teens from substance abuse. The National Center on Addiction and Substance Abuse put out the results in the following pie chart (Fig. 6-12)

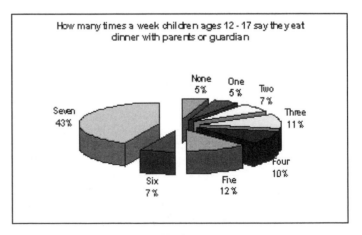

Fig. 6-12.

summarizing the number of times that children aged 12–17 say that they eat dinner with parents or a guardian.

In order to check the figures, a social researcher conducted a similar survey of 1000 teens aged 12–17. She obtained the results shown in Table 6-5. Test the hypothesis that the social researcher obtained a different distribution at $\alpha = 0.05$.

Table 6-5

Times	Number
None	57
One	66
Two	81
Three	101
Four	98
Five	131
Six	81
Seven	385

6. A study was conducted to determine whether shopping on line and age are independent factors. The results are given in Table 6-6. Test that the two factors are independent at $\alpha = 0.05$.

Table 6-6   Buying on line versus age group.

Age group	Number of times you bought on line last year			
	None	1–5	6–10	Over 10
Below 20	25	50	75	150
20 to 40	30	60	70	140
Over 40	95	85	70	50

7.　The following clustered bar chart (Fig. 6-13) contrasts boys' and girls'
TV time. The information appeared in a recent issue of *USA Today*.

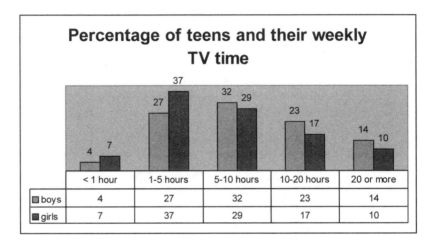

Percentage of teens and their weekly TV time

	< 1 hour	1-5 hours	5-10 hours	10-20 hours	20 or more
boys	4	27	32	23	14
girls	7	37	29	17	10

Fig. 6-13.

Do the times for boys given in Table 6-7 contradict the distribution?
Use $\alpha = 0.05$.

Table 6-7

Times	Frequency
< 1 hour	13
1–5 hours	84
5–10 hours	100
10–20 hours	71
20 or more	32

6-5 Chapter 6 Summary

PROPERTIES OF A MULTINOMIAL EXPERIMENT

1. The experiment has n identical trials.
2. There are K possible outcomes to each trial. These outcomes are referred to as classes, categories, or cells.
3. The probabilities of the K outcomes are represented by $p_1, p_2, \ldots, p_K$ and remain the same from trial to trial. Furthermore, $\sum_{i=1}^{K} p_i = 1$.
4. The trials are independent.
5. The cell counts $n_1, n_2, \ldots, n_K$ are random variables. They represent the number of observations that fall into each of the K categories. They have a multinomial distribution.

CHI-SQUARED GOODNESS-OF-FIT TEST

H_0: Categorical probabilities are specified.
H_a: At least one p_i is not equal to its specified value.

Test Statistic

$$\chi^2 = \sum_{i=1}^{K} \frac{(f_i - e_i)^2}{e_i}$$

where the f_i are the observed values and $e_i = np_i$ are the expected values. The test statistic has an approximate chi-squared distribution with $K - 1$ degrees of freedom when e_i is equal to or greater than 5.

Minitab Solution

Enter the observed values in C1 and the expected values in C2. The pull-down **Calc ⇒ Calculator** gives the Calculator dialog box, which is filled in as $(C1 - C2)**2/C2$. The Expression is stored in C3. The pull-down **Calc ⇒ Column Statistics** requests the sum in C3. This value is the computed test statistic. The p-value is obtained by the pull-down **Calc ⇒ Probability Distribution ⇒ Chi-Square** which gives $P(X < \text{computed test statistic})$. The p-value is then equal to $1 - P(X < \text{computed test statistic})$.

Excel Solution

CHITEST is selected from the paste function dialog box. The expected values are entered into column A and the observed values are entered into column B. Once the observed and expected ranges are supplied to the Chitest dialog box, the *p*-value is given.

CHI-SQUARED TEST OF A CONTINGENCY TABLE

H_0: The two variables are independent.
H_a: The two variables are dependent.

Test Statistic

$$\chi^2 = \sum \frac{(f_{ij} - e_{ij})^2}{e_{ij}}$$

where the sum is over the *r* rows and *c* columns of the contingency table. The expected frequency is

$$e_{ij} = \frac{\text{column } j \text{ total} \times \text{row } i \text{ total}}{\text{sample size}}$$

The test statistic has $(r-1)(c-1)$ degrees of freedom and each expected frequency is equal to or greater than 5.

Minitab Solution

The contingency table is entered in the worksheet. The pull-down **Stat** ⇒ **Tables** ⇒ **Chi-square Test** gives the dialog box, which is filled in. The test statistic value and the *p*-value are produced as output.

Excel Solution

The test statistic is computed using the Excel worksheet. CHIDIST is used to find the *p*-value.

Final Exams and Their Answers

The following examination is a final exam covering the entire six chapters of the book as well as the review chapter. The questions may have anywhere from 0 to 4 answers that are correct. Choose anywhere from none to all four of the letters for the correct answers to the question. A question is correct only if you get the exact answers to it. For example, if problem 1 has (a) and (d) as correct answers, then you must give (a) and (d) as the correct answers. No partial credit is given.

1. Give the level of significance, α, corresponding to the rejection region $Z \geq 2.00$.
   (a) 0.2280   (b) 0.0228   (c) 0.0500   (d) 0.1000

2. Give the level of significance, α, corresponding to the rejection region $|Z| \geq 1.55$.
   (a) 0.5500   (b) 0.0606   (c) 0.1211   (d) 0.1234

3. Give the rejection region in terms of Z for $\alpha = 0.075$ and a lower tailed research hypothesis.
   (a) $Z \le -1.34$ (b) $Z \le -1.38$ (c) $Z \le -1.40$ (d) $Z \le -1.44$

4. Find the p-value that is determined by the following computed test statistic and research hypothesis: $Z = 2.34$, H_a: $\mu \ne \mu_0$.
   (a) 0.0096   (b) 0.0101   (c) 0.0193   (d) 0.0202

5. Find the p-value that is determined by the following computed test statistic and research hypothesis: $Z = -3.05$, H_a: $\mu < \mu_0$.
   (a) 0.0011   (b) 0.0111   (c) 0.0022   (d) 0.0001

6. Find the p-value that is determined by the following computed test statistic and research hypothesis: $Z = 1.89$, H_a: $\mu > \mu_0$.
   (a) 0.0588   (b) 0.1234   (c) 0.0249   (d) 0.0294

7. The null hypothesis is H_0: $\mu = 10$ and the research hypothesis is H_a: $\mu \ne 10$ and $\alpha = 0.01$. For which one of the following 99% confidence intervals for μ would the null hypothesis be rejected?
   (a) (12.1, 15.3)   (b) (8.8, 12.5)   (c) (5.5, 15.5)   (d) (9.9, 10.5)

8. Consider the following Minitab output for a large sample test of the mean.

   One-Sample Z: HA1C

   ```
   Test of mu = 6 vs < 6
   The assumed standard deviation = 0.476
   ```

					95% Upper		
Variable	N	Mean	StDev	SE Mean	Bound	Z	P
HA1C	500	5.51912	0.47608	0.02129	5.55413	−22.59	0.000

   Give the letters of the following statements that are true.
   (a) This is a one-tailed test.
   (b) The computed value of the test statistic is −22.59.
   (c) The sample size is 500.
   (d) The value of μ_0 is 5.51912.

9. A medical study concerning a rare neurological disease is being conducted. The response time to a neurological stimulus is measured for each individual in the study. Study the following Minitab output for the experiment and give the letters of the following statements that are true.

   One-Sample T: time

   ```
   Test of mu = 1.5 vs not = 1.5
   ```

Variable	N	Mean	StDev	SE Mean	95% CI	T	P
time	15	1.97087	0.55233	0.14261	(1.66500, 2.27674)	3.30	0.005

(a) The analysis is for a two-tailed test.

(b) The area to the right of 3.30 under a student t curve having 14 degrees of freedom is 0.005.

(c) The null hypothesis would not be rejected at $\alpha = 0.05$ because 1.5 is not included in the 95% confidence interval (1.66500, 2.27674).

(d) There are 15 individuals in the sample.

10. A poll by American research asked the question "Do you view the media as too liberal?" Study the following Minitab output for the survey and determine which of the following statements are true. (X is the number who responded Yes.)

```
Test of p = 0.4 vs p not = 0.4
Sample  X    N    Sample p  95% CI                  Z-Value  P-Value
1       180  400  0.450000  (0.401247, 0.498753)    2.04     0.041
```

(a) The poll surveyed 400 people and 180 responded "Yes, the media are too liberal."

(b) The analysis is for a one-tailed hypothesis.

(c) The null hypothesis would be rejected at $\alpha = 0.05$.

(d) The area under the standard normal curve beyond 2.04 is 0.041.

11. A telephone survey of 500 individuals over 65 is conducted. The question asked is "When you die, will you have your body cremated?" Study the following Excel output (Fig. 1) and determine which of the following statements are true.

	A	B	C	D	E	F	G	H	I
1	0.25	The sample proportion who will have their body cremated is given by =125/500							
2	0.019365	The standard error of the proportion is given by =SQRT(A1*(1-A1)/500)							
3	1.959961	The 95% confidence interval Z value is given by =NORMSINV(0.975)							
4									
5	0.212046	The lower 95% confidence limit is =A1-A2*A3							
6	0.287954	The upper 95% confidence limit is =A1+A2*A3							
7									
8	2.795085	The computed test statistic is =(A1-0.2)/SQRT(0.2*0.8/500)							
9	0.005189	The computed p-value is =2*(1-NORMSDIST(A8))							
10									

Fig. 1.

(a) The value of p_0 is 0.20.

(b) The value 0.005189 is a one-tailed p-value.

(c) The value of p_0 does not fall in the 95% confidence interval (0.212046, 0.287954).

(d) The sample size is large enough so that $np_0 \geq 5$ and $nq_0 \geq 5$ and the normal approximation to the binomial distribution can be used.

12. A professional golfer claims that the standard deviation of her golf scores is less than 5. A newspaper sports writer randomly selects 12 eighteen-hole rounds and records the golfer's scores. The scores are shown in Table 1. The Excel analysis is shown in Fig. 2.

Table 1   Golf scores for golf pro.

70	73	70
72	74	70
74	76	75
80	74	80

	A	B	C	D	E	F	G	H	I
1	70		3.437758	The standard deviation is given by =STDEV(A1:A12)					
2	72		5.2	The computed test statistic is =11*C1^2/25					
3	74								
4	80		0.078905	The p-value is =1-CHIDIST(C2,11)					
5	73			Note that CHIDIST(5.2,11) gives the area to the right of 5.2.					
6	74			Since we have a lower tailed test the p-value is =1-CHIDIST(C2,11)					
7	76								
8	74								
9	70			Since the p-value is > 0.05, do not reject the null hypothesis					
10	70								
11	75								
12	80								
13									
14									

Fig. 2.

Which of the following statements are true?

(a) S^2 for the sample is equal to 3.43776.

(b) The null hypothesis should not be rejected at $\alpha = 0.05$.

(c) The area to the left of 5.2 under the chi-squared curve with 11 degrees of freedom is 0.078905.

(d) This is an upper-tailed test.

13. A major cause of accidents is the differing speeds of vehicles on the roadway. A sample of interstate speeds was obtained and is given in Table 2. Excel was used to set a 95% confidence interval on the standard deviation of the speeds on the interstate. The Excel output (Fig. 3) follows the table of speeds. Refer to the Excel analysis and the statements below to determine the true statements.

Table 2 Speeds of 100 vehicles on Interstate 80 near Omaha, Nebraska.

75	75	75	60	85	60	70	85	70	75
50	75	60	75	65	70	75	75	80	80
55	65	80	85	60	65	80	75	85	85
50	70	55	60	55	55	70	55	80	55
75	50	75	50	70	60	65	75	80	50
70	75	60	75	65	65	60	85	85	60
75	75	75	75	55	55	75	50	60	60
80	50	70	70	75	70	55	85	75	80
75	60	75	75	85	80	65	70	50	60
75	75	50	60	80	60	50	75	75	65

(a) $S^2 = 10.64818$ for the 100 numbers in Table 2.
(b) A 95% confidence interval on σ^2 is (87.40719, 153.0102).
(c) A 95% confidence interval on σ is (9.349181, 12.36973).
(d) There is 99% of the area under the chi-squared curve with 99 degrees of freedom between 73.3611 and 128.4219.

14. A hypothesis is being tested at $\alpha = 0.05$. For which of the following p-values would the null hypothesis be rejected.
(a) p-value $= 0.005$ (b) p-value $= 0.14$ (c) p-value $= 0.024$
(d) p-value $= 0.34$

15. Statistical inference is concerned with which of the following?
(a) Describing samples.
(b) Describing populations.

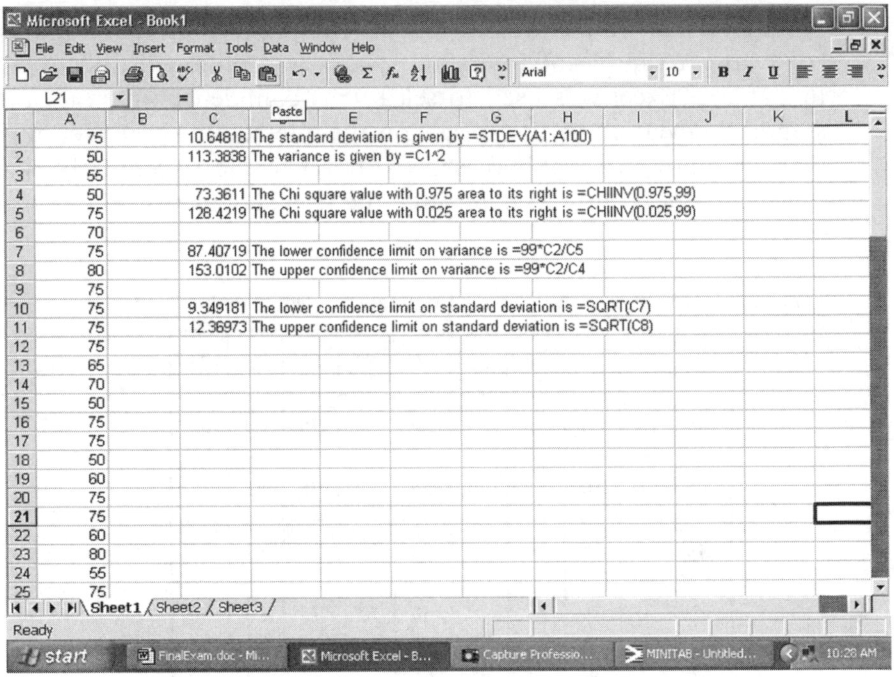

Fig. 3.

(c) Making inferences from samples to populations.

(d) Making inferences from populations to samples.

16. When performing a two independent samples t-test, it is assumed that:

(a) The samples are independent of one another.

(b) The samples come from normally distributed populations.

(c) The samples come from a t-distribution with $n_1 + n_2 - 2$ degrees of freedom.

(d) The samples come from the same population.

17. Suppose you wished to compare the percent of people who are 30 or below who attend church regularly with those who are 65 and above who attend regularly. Five hundred from each group were asked the question "In a typical week, do you attend church?" Sample 1 is from the 30 or below population. The following is a Minitab output analysis of the data.

Sample	X	N	Sample p
1	200	500	0.400000
2	275	500	0.550000

Difference = p(1) − p(2)

```
Estimate for difference: -0.15
95% CI for difference: (-0.211200, -0.0888001)
Test for difference = 0 (vs not = 0): Z = -4.80 P-Value = 0.000
```

Which of the following statements are true?
(a) Fifteen percent more in the 30 or below group attended church in a typical week than did in the 65 and above group.
(b) The value of the test statistic is -4.80.
(c) The area to the left of $z = -4.80$ plus the area to the right of $z = 4.80$ equals the p-value.
(d) The test statistic has a standard normal distribution.

18. Researchers monitored 46 men and women with high cholesterol: 16 ate a prepackaged vegetarian diet for one month, 16 consumed a low-fat diet, and 14 ate a low-fat diet and took 20 milligrams of lovastatin every day. The response measured was the reduction in cholesterol. The results of a one-way ANOVA were as follows:

Source	DF	SS	MS	F	P
Factor	2	2076.57	1038.29	154.70	0.000
Error	43	288.59	6.71		
Total	45	2365.16			

S = 2.591   R-Sq = 87.80%   R-Sq(adj) = 87.23%

```
Individual 95% CIs For Mean Based on
                                Pooled StDev
Level        N    Mean    StDev   - - - + - - - - + - - - + - - - + - - -
Vegetarian   16   23.960  1.705                           (- - * - -)
Low-fat      16   10.324  2.069   (- - * - -)
Lovastatin   14   24.918  3.729                               (- - * - -)
                                  - - - + - - - + - - - + - - - + - - -
                                  10.0     15.0    20.0    25.0
```

Which of the following statements are true?
(a) The p-value was computed assuming there was a mean difference in the three treatments.
(b) It appears that the vegetarian diet and the diet that included lovastatin produced a similar lowering of cholesterol.
(c) There appears to be no difference in the three treatments.
(d) The analysis of variance assumes equal population variances and normality of the response variable.

19. Researchers monitored 46 men and women with high cholesterol: 16 ate a prepackaged vegetarian diet for one month, 16 consumed a low-fat diet, and 14 ate a low-fat diet and took 20 milligrams of lovastatin every day. The response measured was the reduction in

cholesterol. The results of a Kruskal–Wallis procedure are as follows. Which of the four statements are true?

Kruskal–Wallis Test on reduction

treatment	N	Median	Ave Rank	Z
1	16	23.89	30.3	2.51
2	16	10.29	8.5	−5.54
3	14	24.67	32.9	3.13
Overall	46		23.5	

H = 30.91   DF = 2   P = 0.000

(a) The Kruskal–Wallis procedure assumes equal population variances and normality of the response variable.

(b) $H = 30.91$ is the value of the test statistic and it has an approximate F-distribution.

(c) $H = 12/(n(n+1)) \sum R_i^2/n_i - 3(n+1)$. In the formula, $n = 46$, $n_1 = 16$, $n_2 = 16$, $n_3 = 14$, $R_1 = 485$, $R_2 = 136$, and $R_3 = 460$.

(d) The sum of the ranks is 1081.

20. Researchers monitored 30 men and women with high cholesterol: 16 ate a prepackaged vegetarian diet for one month and 14 ate a low-fat diet and took 20 milligrams of lovastatin every day. The response measured was the reduction in cholesterol. The Excel results of a two sample t-test were as follows:

vegetarian	lovastatin	t-Test: Two-Sample Assuming Equal Variances		
25	19			
25	27		vegetarian	lovastatin
23	24	Mean	24.0625	24.9286
21	23	Variance	2.99583333	13.9176
24	32	Observations	16	14
26	24	Pooled Variance	8.06664541	
23	28	Hypothesized Mean Difference	0	
27	28	Df	28	
24	22	t Stat	−0.83324096	
22	30	P(T<=t) one-tail	0.20588025	
25	25	t Critical one-tail	1.70113026	
22	20	P(T<=t) two-tail	0.41176051	
27	25	t Critical two-tail	2.04840944	
24	22			
24				
23				

Which of the following statements are true?
(a) The value of the calculated test statistic is −0.833.
(b) The area to the left of −0.833 is the one-tailed p-value and equals 0.206.
(c) The area to the left of −0.833 plus the area to the right of 0.833 is the two tailed p-value and is equal to 0.411.
(d) The test statistic has a standard normal distribution.

21. Researchers monitored 30 men and women with high cholesterol: 16 ate a prepackaged vegetarian diet for one month and 14 ate a low-fat diet and took 20 milligrams of lovastatin every day. The response measured was the reduction in cholesterol. The Minitab output for a Mann–Whitney test is given below:

Mann–Whitney Test and CI: vegetarian, lovastatin

	N	Median
vegetarian	16	23.891
lovastatin	14	24.674

Point estimate for ETA1-ETA2 is −0.750
95.2 Percent CI for ETA1-ETA2 is (−3.155,1.513)
W = 229.0
Test of ETA1 = ETA2 vs ETA1 not = ETA2 is significant at 0.4419

Which of the following statements are true?
(a) When the two samples are jointly ranked, one has a sum of ranks equal to 229 and the other has a sum of ranks equal to 236.
(b) When the data are replaced by ranks, the sum of all the ranks is 500.
(c) The two tailed p-value equals 0.4419.
(d) The one tailed p-value equals 0.22095.

22. Twenty adult males were randomly selected and their weights and heights were determined. A plot of the data is shown in the Fig. 4. Which of the following statements are true?
(a) The two variables are directly related.
(b) The correlation coefficient r is a positive number.
(c) The two variables are not related.
(d) Generally, as the height increases, the weight increases.

23. An article appeared in a recent issue of *USA Today* entitled "Lighter cars mean more deaths." The following table appeared in the article. The data were taken from the National Highway Traffic Safety Administration. Table 3 gives the driver fatalities per billion vehicle miles from 1996 to 2000 in 1996 to 1999 models. The statistical

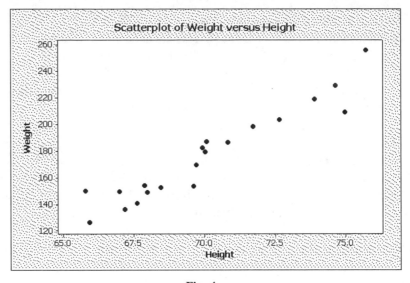

Fig. 4.

Table 3 Lighter cars mean more deaths.

Size of car	Fatality rate
Very small cars	11.56
Small cars	7.85
Compact pickups	6.82
Midsize SUVs	6.73
Small SUVs	5.68
Midsize cars	5.26
Large pickups	4.07
Large SUVs	3.79
Large cars	3.30
Minivans	2.76

concept that is illustrated by the article is:
(a) The size of the car and the fatality rate are directly related.
(b) The size of the car and the fatality rate are not related.
(c) The size of the car and the fatality rate are inversely related.
(d) The size of the car and the fatality rate are polynomially related.

24. In a recent *USA Today* article entitled "Churchgoing increases with age" the following information appeared. Table 4 gives the average

Table 4 Churchgoing increases with age.

Age group	Percent
20s	31%
30s	42%
40s	47%
50s	48%
60+	53%

percentage for each age group who attended church services. The statistical concept that is illustrated by the article is:
(a) Church attendance and age are directly related.
(b) Church attendance and age are inversely related.
(c) Church attendance and age are quadratically related.
(d) Church attendance and age are not related.

25. A study was done in which 25 teens were asked to give the number of hours per week they spent watching TV and the number of hours per week they spent on the Internet. Differences were formed where *Difference* = *number of hours spent on the Internet* − *number of hours spent watching TV*. The research hypothesis was H_a: difference > 0. Which of the following statements are true?
(a) The design here is the paired design.
(b) The research hypothesis is that the teens spend less time on the Internet than watching TV.
(c) The research hypothesis is a one-tailed hypothesis.
(d) The design here is the independent samples design.

26. A study was done in which 25 teens were asked to give the number of hours per week they spent watching TV and the number of

hours per week they spent on the Internet. Differences were formed where *Difference = number of hours spent on the Internet – number of hours spent watching TV*. The research hypothesis was H_a: Difference > 0. The Minitab output is shown below.

```
Paired T for Internet - TV
                N        Mean          StDev         SE Mean
Internet        25       7.88000       2.69753       0.53951
TV              25       6.76000       2.35018       0.47004
Difference      25       1.12000       2.99054       0.59811

95% lower bound for mean difference: 0.09671
T-Test of mean difference = 0 (vs > 0): T-Value = 1.87
P-Value=0.037
```

Which of the following statements are true?
(a) For the analysis to be valid, the differences must be normally distributed.
(b) The test statistic has a *t*-distribution with 25 degrees of freedom.
(c) The null hypothesis in the analysis above is rejected for $\alpha = 0.05$.
(d) The area to the right of 1.87 on the student-*t* curve with 25 degrees of freedom is 0.037.

27. A study was done in which 25 teens were asked to give the number of hours per week they spent watching TV and the number of hours per week they spent on the Internet. Differences were formed where *Difference = number of hours spent on the Internet – number of hours spent watching TV*. The research hypothesis was H_a: Difference > 0. The Minitab output for the 1-sample Wilcoxon analysis is shown below.

Wilcoxon Signed Rank Test: Difference
```
Test of median = 0.000000 versus median > 0.000000
                         N
                        for      Wilcoxon                 Estimated
            N           Test     Statistic    P           Median
Difference  25          21       159.0        0.068       1.000
```

Which of the following statements are true?
(a) At $\alpha = 0.05$, the null hypothesis would not be rejected.
(b) There are four differences that are zero.
(c) When the absolute differences are ranked, the sum of the ranks is 231.
(d) This is the test to do if the differences are not normally distributed.

28. A company wished to compare its inspectors on the night shift with those on the day shift. Management presented 100 items for inspection to the inspectors on the night and the day shift. The quality of each item was known to the manager. The eight inspectors on each shift evaluated the 100 items. The number of inspection errors was recorded for each of the 16 inspectors. Table 5 lists the number of inspection errors made by each inspector. The inspection errors were like type 1 and type 2 errors: a non-defective item could be classified as defective or a defective item could be classified as non-defective.

Table 5 Inspection errors made by the day shift and the night shift.

Day shift		Night shift	
10	11	4	5
4	4	4	6
7	4	5	4
8	5	5	6

The null hypothesis $H_0: \sigma_1^2 = \sigma_2^2$ was to be tested against $H_a: \sigma_1^2 \neq \sigma_2^2$. Which of the following statements are true?
(a) **Stat $\Rightarrow$ Basic Statistics $\Rightarrow$ 2 Variances** is the Minitab pull-down to perform the test.
(b) The Excel paste function **FTEST** is the Excel function that performs the test.
(c) The test assumes that you have two independent samples from populations that are normally distributed.
(d) The test statistic that the test is based upon is the standard normal.

29. A company wished to compare its inspectors on the night shift with those on the day shift. Management presented 100 items for inspection to the inspectors on the night and the day shift. The quality of each item was known to the manager. The eight inspectors on each shift evaluated the 100 items. The number of inspection errors was recorded for each of the 16 inspectors. Table 5 lists the number of

inspection errors made by each inspector. The inspection errors were like type 1 and type 2 errors: a non-defective item could be classified as defective or a defective item could be classified as non-defective.

The null hypothesis H_0: $\sigma_1^2 = \sigma_2^2$ is to be tested against the research hypothesis H_a: $\sigma_1^2 \neq \sigma_2^2$. The Minitab output for the analysis is as follows.

Test for Equal Variances: day, night

```
95% Bonferroni confidence intervals for standard deviations
             N       Lower       StDev       Upper
day          8     1.76752     2.82527     6.47123
night        8     0.52209     0.83452     1.91146

F-Test (normal distribution)
Test statistic = 11.46, p-value = 0.005
Levene's Test (any continuous distribution)
Test statistic = 10.89, p-value = 0.005
```

Which of the following statements are true?
(a) The night group is less variable than the day group.
(b) The test is based on the F-distribution with 7 and 7 degrees of freedom.
(c) At $\alpha = 0.05$, we would conclude that the two groups are not equally variable.
(d) The F-test is based on the assumption that both samples come from normally distributed populations.

30. A company wished to compare its inspectors on the night shift with those on the day shift. Management presented 100 items for inspection to the inspectors on the night and the day shift. The quality of each item was known to the manager. The eight inspectors on each shift evaluated the 100 items. The number of inspection errors was recorded for each of the 16 inspectors. Table 5 lists the number of inspection errors made by each inspector. The inspection errors were like type 1 and type 2 errors: a non-defective item could be classified as defective or a defective item could be classified as non-defective.

The null hypothesis H_0: $\sigma_1^2 = \sigma_2^2$ is to be tested against the research hypothesis H_a: $\sigma_1^2 \neq \sigma_2^2$. The Excel output follows.

	A	B	C	D	E	F
1	day	night	F-Test Two-Sample for Variances			
2	10	4				
3	4	4		day	night	
4	7	5	Mean	6.625	4.875	
5	8	5	Variance	7.9821429	0.696429	
6	11	5	Observations	8	8	
7	4	6	df	7	7	
8	4	4	F	11.461538		
9	5	6	P(F<=f) one-tail	0.0023218		
10			F Critical one-tail	3.7870507		
11						

Which of the following statements are true?

(a) The night shift makes more inspection errors on the average than the day shift.

(b) There is more variability among the day shift than the night shift.

(c) The p-value is $2(0.002321809) = 0.005$.

(d) At $\alpha = 0.05$, we conclude that there is no difference in the variability of the day and the night shifts.

31. Suppose you wish to test the null hypothesis that $H_0: P_1 = P_2$ against the alternative $H_a: P_1 \neq P_2$. In which of the following cases are the sample sizes large enough to assure the normal approximation is valid?

(a) $n_1 = 25$, $x_1 = 4$, $n_2 = 25$, $x_2 = 7$.

(b) $n_1 = 100$, $x_1 = 3$, $n_2 = 100$, $x_2 = 5$.

(c) $n_1 = 75$, $x_1 = 2$, $n_2 = 75$, $x_2 = 7$.

(d) $n_1 = 250$, $x_1 = 24$, $n_2 = 250$ $x_2 = 17$.

32. Which of the following statements about block designs are true?

(a) The test statistic for testing for block differences or treatment differences is the chi-squared statistic.

(b) The purpose of blocking is to reduce the variation caused by differences between experimental units.

(c) In a block design, the total variation is partitioned into block variation, treatment variation, and error variation.

(d) When the response variable is not normally distributed or the population variances are not equal, the Kruskal–Wallis analysis of a block design should be performed.

33. The Fudge cookie company designs an experiment to compare the sales produced by arranging its cookies in one of three displays. The cookies are sold in several different supermarkets consisting of several different stores each. The three displays are set up in three stores for each of five different supermarkets. The three displays

(treatments) are randomly assigned to three stores randomly selected within each supermarket (block). The sales in dollars made during a given week for each store within each supermarket are given in Table 6. The Minitab parametric block analysis is shown below.

Table 6   Dollar sales of Fudge cookies at 3 stores within 5 supermarkets.

Supermarket (blocks)	Display 1 Treatment 1	Display 2 Treatment 2	Display 3 Treatment 3
Abelbakers	2340	1543	1987
Low Vee	3456	3005	3245
Alberts	1457	1200	1356
Food for more	3400	2980	3246
Hinky Dinky	678	546	600

Two-way ANOVA: sales versus display, supermarket

```
Source        DF    SS          MS         F         P
display        2      425431    212715    13.29     0.003
supermarket    4    16023484   4005871   250.32     0.000
Error          8      128022     16003
Total         14    16576937

S = 126.5      R-Sq = 99.23%    R-Sq(adj) = 98.65%
                                Individual 95% CIs For Mean Based on
                                Pooled StDev

display    Mean        - - - + - - - + - - - + - - - + - - -
1          2266.2                               (- - * - -)
2          1854.8      (- - * - -)
3          2086.8                   (- - * - -)
                       - - - + - - - + - - - + - - - + - - -
                       1800    2000    2200    2400
```

```
                    Individual 95% CIs For Mean Based on
                    Pooled StDev
supermarket         Mean      - - - + - - - + - - - + - - - + - - -
1                   1956.67                 (- * -)
2                   3235.33                               (- * -)
3                   1337.67            (- * -)
4                   3208.67                               (- * -)
5                    608.00   (- * -)
                              - - - + - - - + - - - + - - - + - - -
                               800    1600    2400    3200
```

Which of the following statements are true?
(a) At $\alpha = 0.05$, there is a difference in sales due to displays.
(b) At $\alpha = 0.05$, there is a difference in sales due to supermarkets.
(c) Display 1 outsold Display 2 by $411.40.
(d) A one-way ANOVA would have been just as effective in finding display differences as was the block design.

34. The Fudge cookie company designs an experiment to compare the sales produced by arranging its cookies in one of three displays. The cookies are sold in several different supermarkets consisting of several different stores each. The three displays are set up in three stores for each of five different supermarkets. The three displays (treatments) are randomly assigned to three stores randomly selected within each supermarket (block). The sales in dollars made during a given week for each store within each supermarket are given in Table 6. The Excel parametric block analysis is shown in Fig. 5.

Which of the following statements are true?
(a) The row sums given in the summary are the total sales for the supermarkets in the study.
(b) The column variances are the variances for the displays.
(c) The F critical value for rows is the point on the F-distribution curve having 4 and 8 degrees of freedom where there is only 5% of the area beyond it.
(d) The F critical value for columns is the point on the F-distribution curve having 4 and 8 degrees of freedom where there is only 5% of the area beyond it.

35. The Fudge cookie company designs an experiment to compare the sales produced by arranging its cookies in one of three displays. The cookies are sold in several different supermarkets consisting of several different stores each. The three displays are set up in three stores for each of five different supermarkets. The three displays (treatments) are randomly assigned to three stores randomly selected

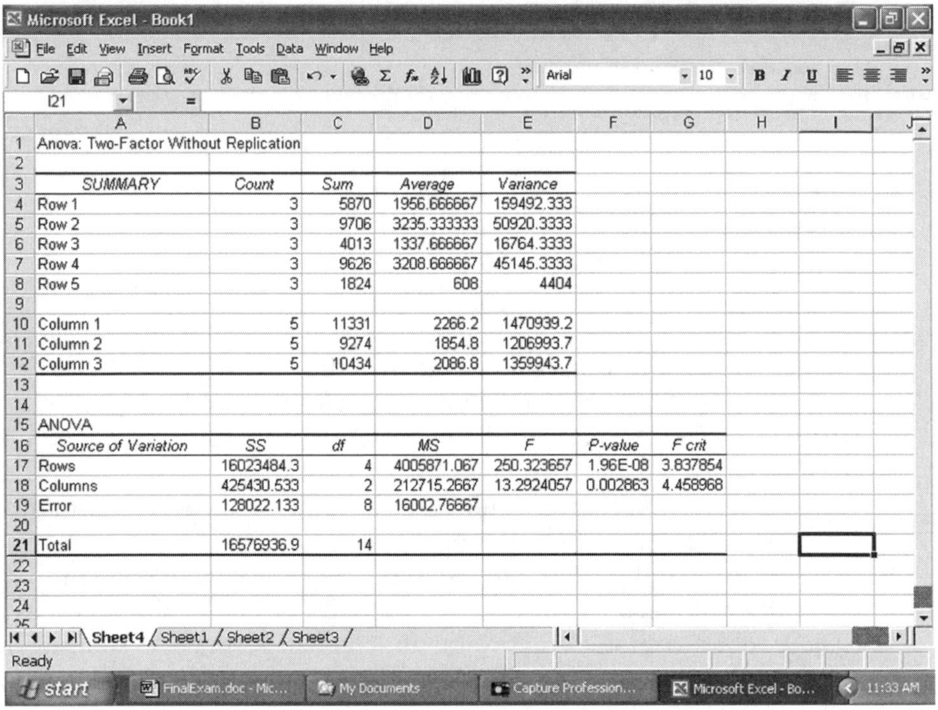

	A	B	C	D	E	F	G
1	Anova: Two-Factor Without Replication						
2							
3	SUMMARY	Count	Sum	Average	Variance		
4	Row 1	3	5870	1956.666667	159492.333		
5	Row 2	3	9706	3235.333333	50920.3333		
6	Row 3	3	4013	1337.666667	16764.3333		
7	Row 4	3	9626	3208.666667	45145.3333		
8	Row 5	3	1824	608	4404		
9							
10	Column 1	5	11331	2266.2	1470939.2		
11	Column 2	5	9274	1854.8	1206993.7		
12	Column 3	5	10434	2086.8	1359943.7		
13							
14							
15	ANOVA						
16	Source of Variation	SS	df	MS	F	P-value	F crit
17	Rows	16023484.3	4	4005871.067	250.323657	1.96E-08	3.837854
18	Columns	425430.533	2	212715.2667	13.2924057	0.002863	4.458968
19	Error	128022.133	8	16002.76667			
20							
21	Total	16576936.9	14				

Fig. 5.

within each supermarket (block). The sales in dollars made during a given week for each store within each supermarket are given in Table 6. The Minitab nonparametric analysis of a block design is shown below.

Friedman Test: sales versus display blocked by supermarket

```
S = 10.00      DF = 2        P = 0.007
                                            Sum
                                            of
display        N             Est Median     Ranks
1              5             2149.7         15.0
2              5             1738.3          5.0
3              5             1987.0         10.0
Grand median = 1958.3
```

Which of the following statements are true?

(a) The test statistic for the Friedman test has an F-distribution.

(b) The estimated medians and the grand median go in to make up the computation of the nonparametric test statistic.

(c) The sum of ranks for display 2 is 5 because display 2 is last in sales at each of the 5 supermarkets.

(d) The expression $F_r = 12/(bp(p+1)) \sum R_i^2 - 3b(p+1)$ is the same as S.

36. Table 7 gives the in-state and out-state tuition fees for nine randomly selected public universities for 2003. The Pearson correlation coefficient between the two variables is:
    (a) 0.22   (b) 0.43   (c) 0.66   (d) 0.91

Table 7   In-state and out-state tuitions for nine public universities.

Out-state	4194	6812	7010	4547	7975	4104	2826	6739	5051
In-state	11354	17596	18046	11227	24778	11475	11313	16603	12131

37. Table 7 gives the in-state and out-state tuition fees for nine randomly selected public universities for 2003.

```
The regression equation is
out-state = 1339 + 2.49 in-state
Predictor    Coef        SE Coef      T       P
Constant     1339        2443         0.55    0.601
in-state        2.4864      0.4281    5.81    0.001

S = 2078.34 R-Sq = 82.8% R-Sq(adj) = 80.4%

Analysis of Variance

Source           DF   SS          MS         F       P
Regression       1    145714072   145714072  33.73   0.001
Residual Error   7     30236472     4319496
Total            8    175950544
```

Consult the Minitab output and tell which of the following statements are true.

(a) The p-value for the test of the hypothesis H_0: $\beta_1 = 0$ against the alternative H_a: $\beta_1 \neq 0$ is 0.601.

(b) The least-squares line connecting in-state and out-state tuition is out-state $= 1339 + 2.49$ in-state.

(c) For every \$1 increase in in-state tuition, out-state tuition increases \$2.49.

(d) The correlation coefficient between in-state tuition and out-state tuition is $\sqrt{0.828}$.

38. Refer to the Minitab output associated with problem 37. Which of the following statements are true.
    (a) $\sqrt{80.4}\%$ of the variation in out-state tuition fees can be accounted for by the straight line model out-state $= 1339 + 2.49$ in-state.
    (b) 95% of the values of out-state tuition fees fall within $2s = \$4156.68$ of the line connecting in-state tuition with out-state tuition.
    (c) The slope of the line of best fit is 2.49.
    (d) The predicted out-state tuition fee for a public university that has an in-state tuition fee of $5000 is $15,000.

39. In a recent issue of *USA Today* the worldwide growth of spam messages sent daily was given for the years 1999 through 2003. Table 8 shows the growth of spam messages. The least-squares fit is given below.

```
Row            year         billions
1              1            1.0
2              2            2.3
3              3            4.0
4              4            5.6
5              5            7.3
```

```
The regression equation is
billions = - 0.730 + 1.59 year
```

Table 8   Spam messages.

Year	Billions
1999	1.0
2000	2.3
2001	4.0
2002	5.6
2003	7.3

If the variable year is coded 1, 2, 3, 4, and 5 and it is assumed that the linear growth will continue for 2004, what is the predicted daily

world wide spam in billions sent for 2004?
(a) 8.2   (b) 8.4   (c) 8.6   (d) 8.8

40. Table 9 gives a data set from a research study involving the relationship between computer science achievement and several other

Table 9   Multiple regression study: Response = CSFINAL as a function of 6 independent variables.

HSAP	CCAP	CUC	PECS	KSW	MC	CSFINAL
3	3	2	2	4	28	19
1	1	2	1	16	35	19
3	2	5	2	20	42	24
3	4	1	2	13	41	36
1	2	5	2	22	44	27
2	2	3	2	21	42	26
1	1	5	1	15	36	25
1	1	5	1	20	44	28
2	2	3	2	19	39	17
2	2	1	1	16	36	27
3	2	3	1	18	40	21
2	3	4	1	17	40	24
3	1	3	2	8	33	18
3	4	1	1	6	27	18
2	2	5	1	5	44	14
2	1	5	1	20	45	28
2	2	3	1	18	41	21

(Continued)

Table 9 Continued.

HSAP	CCAP	CUC	PECS	KSW	MC	CSFINAL
3	3	1	1	11	31	22
2	4	1	1	19	41	20
2	2	5	1	14	43	21
2	2	4	2	22	43	35
2	1	3	2	13	31	22
3	3	3	2	16	39	18
2	2	2	1	6	26	14
2	2	4	1	7	34	16
2	2	3	2	17	41	27
3	3	2	1	7	37	29
2	2	4	2	21	46	29
2	1	4	1	20	41	28
1	2	3	2	23	48	29
2	2	1	1	15	40	25
2	3	5	1	16	42	17
2	1	2	1	14	43	20
1	2	3	2	16	45	26
1	2	5	1	21	45	28
1	1	5	2	25	47	33
1	1	1	1	18	42	31
2	2	2	1	16	38	27

variables (Konvalina, Stephens, and Wileman, early 1980s). The variables are defined as follows:

- HSAP is high school academic performance: 1=excellent, 2=above average, 3=average or below
- CCAP is current college academic performance: 1=excellent, 2=above average, 3=average or below, 4=no past college experience
- CUC is current university classification: 1=freshman, 2=sophomore, 3=junior, 4=senior, 5=other
- PECS is previous education in computer science: 1=no computer courses in high school, 2=at least one computer course in high school
- KSW=computer science aptitude test score, ranging from 0 to 25
- MC=math competency score, ranging from 0 to 48
- CSFINAL=computer science achievement score, ranging from 0 to 40

The Minitab multiple regression output is as follows.

Regression Analysis: CSFINAL versus HSAP, CCAP, CUC, PECS, KSW, MC

```
The regression equation is
CSFINAL = 6.18 − 0.12 HSAP − 0.25 CCAP − 1.11 CUC + 1.66 PECS +
          0.392 KSW + 0.343 MC
```

Predictor	Coef	SE Coef	T	P
Constant	6.178	7.846	0.79	0.437
HSAP	−0.118	1.497	−0.08	0.938
CCAP	−0.255	1.090	−0.23	0.817
CUC	−1.1078	0.6397	−1.73	0.093
PECS	1.661	1.639	1.01	0.319
KSW	0.3915	0.2264	1.73	0.094
MC	0.3426	0.2266	1.51	0.141

```
S = 4.61416   R-Sq = 43.4%          R-Sq(adj) = 32.5%
```

Analysis of Variance

Source	DF	SS	MS	F	P
Regression	6	506.76	84.46	3.97	0.005
Residual Error	31	660.01	21.29		
Total	37	1166.76			

Which of the following statements are true?
(a) A 95% confidence interval for β_6 is $0.3426 \pm 1.96(0.2266)$.
(b) The coefficient of determination is 43.4%.
(c) The adjusted coefficient of determination is 32.5%.
(d) At $\alpha = 0.05$ and for a one-tailed research hypothesis CUC and KSW have a significant effect on CSFINAL.

41. From the data of problem 40, the Minitab package is used to find the 95% prediction interval and the 95% confidence interval for the values of the predictors for new observations shown below.

```
Predicted Values for New Observations
New
Obs        Fit       SE Fit    95% CI            95% PI
1          26.231    1.880     (22.397, 30.065)  (16.069, 36.393)

Values of Predictors for New Observations
New
Obs        HSAP      CCAP      CUC      PECS     KSW      MC
1          1.00      1.00      4.00     2.00     20.0     40.0
```

Which of the following statements are true?
(a) A point estimate of the CSFINAL score for a student with HSAP$=1$, CCAP$=1$, CUC$=4$, PECS$=2$, who scored 20 on the KSW, and who scored 40 on the MC, is 26.231.
(b) A point estimate of the CSFINAL mean score for all students with HSAP$=1$, CCAP$=1$, CUC$=4$, PECS$=2$, who scored 20 on the KSW, and who scored 40 on the MC, is 26.231.
(c) A 95% prediction interval for a student with HSAP$=1$, CCAP$=1$, CUC$=4$, PECS$=2$, who scored 20 on the KSW, and who scored 40 on the MC, is (16.069, 36.393).
(d) A 95% confidence interval of the CSFINAL mean score for all students with HSAP$=1$, CCAP$=1$, CUC$=4$, PECS$=2$, who scored 20 on the KSW, and who scored 40 on the MC, is (22.397, 30.065).

42. The Excel analysis of the data in problem 40 is shown in Fig. 6. Refer to the Excel output for the multiple regression analysis to answer the following. Which of the following statements are true?
(a) The multiple coefficient of determination is 0.659.
(b) A 95% confidence interval for the variable KSW is -0.070 to 0.853.
(c) The p-value $=0.005$ in the ANOVA table is a measure of the overall utility of the model.
(d) The equation $R\text{-sq(adj)} = 1 - [(n-1)/(n-(k+1))](1 - R\text{-sq})$,

SUMMARY OUTPUT

Regression Statistics	
Multiple R	0.659035267
R Square	0.434327483
Adjusted R Square	0.32484248
Standard Error	4.614164211
Observations	38

ANOVA

	df	SS	MS	F	Significance F
Regression	6	506.7573056	84.45955093	3.967004	0.0046551
Residual	31	660.0058523	21.29051137		
Total	37	1166.763158			

	Coefficients	Standard Error	t Stat	P-value	Lower 95%	Upper 95%
Intercept	6.177944938	7.84643449	0.787356977	0.437052	-9.824972635	22.18086
HSAP	-0.11824595	1.496558173	-0.079011928	0.937531	-3.170498166	2.934006
CCAP	-0.25456318	1.089833582	-0.233579862	0.816847	-2.477294661	1.968168
CUC	-1.10779894	0.639722543	-1.731686577	0.093271	-2.412522396	0.196925
PECS	1.661039031	1.639214042	1.013314301	0.318754	-1.682161914	5.00424
KSW	0.391533131	0.226361396	1.729681552	0.093634	-0.070134238	0.8532
MC	0.34260767	0.226552378	1.51226693	0.140595	-0.11944921	0.804665

Fig. 6.

where n is 38 and $k = 6$, connects the coefficient of determination and the adjusted coefficient of determination.

43. Refer to the data in problem 40. A stepwise regression was performed with CSFINAL as the dependent variable and the other six variables as independent variables. The results are shown below.

Stepwise Regression: CSFINAL versus HSAP, CCAP, CUC, PECS, KSW, MC

```
Forward selection. Alpha-to-Enter: 0.25

Response is CSFINAL on 6 predictors, with N = 38
Step          1        2         3
Constant   14.478   15.618    7.130

KSW         0.60      0.67     0.45
T-Value     4.37      4.56     2.23
P-Value     0.000     0.000    0.032

CUC                  −0.69   −1.02
T-Value              −1.26   −1.74
P-Value               0.218    0.090
```

MC			0.33
T-Value			1.52
P-Value			0.139
S	4.60	4.56	4.48
R-Sq	34.68	37.50	41.45
R-Sq(adj)	32.87	33.93	36.29
Mallows C-p	1.8	2.3	2.1

Which of the following statements are true?

(a) The equation of the best regression model with one independent variable is CSFINAL $= 14.478 + 0.60$ KSW.

(b) The equation of the best regression model with two independent variables is CSFINAL $= 15.618 + 0.67$ KSW $- 0.69$ CUC.

(c) The equation in step 3 accounts for 36.29% (adjusted) of the variation in the CSFINAL grades.

(d) This type of regression is called forward stepwise.

44. A taste test was conducted in which 20 people were asked to taste both Coke and Pepsi. They tasted both and stated which they preferred. The null hypothesis is that you cannot tell the difference between the two and the research hypothesis is you can tell the difference. The sign test may be used to tell whether there is a real difference. Let $X =$ the number in the 20 that prefer Coke. Under the null hypothesis, X is binomial with $n = 20$ and $p = 0.5$. Suppose, when the test was conducted, 15 preferred Coke. The p-value $= 2P(X \geq 15)$. Using the binomial distribution, the p-value is $2(0.0059) = 0.0118$.

A normal approximation is as follows: $z = (x - np)/\sqrt{npq} = (15 - 20(0.5))/\sqrt{20(0.5)(0.5)} = 2.24$. The p-value using the normal approximation is $2P(Z > 2.24) = 2(0.0125) = 0.025$.

Using the chi-squared approximation at the beginning of Chapter 6, $\chi^2 = (n_1 - np_1)^2/(np_1) + (n_2 - np_2)^2/(np_2) = (15 - 10)^2/10 + (5 - 10)^2/10 = 2.5 + 2.5 = 5.0$. The area to the right of 5.0 under the chi-squared curve is 0.0253.

Which of the following statements are true?

(a) At $\alpha = 0.05$, the null hypothesis would be rejected no matter which of the three methods is used.

(b) The most accurate of the three approaches is the one that uses the Z variable.

(c) The most accurate of the three approaches is the one that uses the chi-squared approximation.

(d) There are three test statistics used to test the same hypothesis: the binomial X, the standard normal Z, and the chi-squared χ^2. Regardless of the sample size, the three methods are equally accurate.

45. Thirty people participate in a taste test. They are given two coffees and are asked which one tastes the best. The null hypothesis is H_0: no difference in the two coffees, versus H_a: there is a difference. Alpha is chosen to be 0.05. When the null hypothesis is true, the taste test may be viewed as a binomial experiment having 30 trials with $p = 0.5$. Figure 7 shows a partial Excel output for the cumulative binomial distribution.

	A	B
		=BINOMDIST(A1,30,0.5,1)
1	0	9.31323E-10
2	1	2.8871E-08
3	2	4.33996E-07
4	3	4.21517E-06
5	4	2.97381E-05
6	5	0.000162457
7	6	0.000715453
8	7	0.00261144
9	8	0.008062401
10	9	0.021386973
11	10	0.049368573
12	11	0.100244211
13	12	0.180797304
14	13	0.292332356
15	14	0.427767776
16	15	0.572232224
17	16	0.707667644
18	17	0.819202696
19	18	0.899755789
20	19	0.950631427
21	20	0.978613027
22	21	0.991937599
23	22	0.99738856
24	23	0.999284547
25	24	0.999837543

Fig. 7.

The rejection region for the test is:
(a) $0 \leq X \leq 9$ or $21 \leq X \leq 30$, where X = the number who prefer brand A over brand B
(b) $0 \leq X \leq 10$.
(c) $20 \leq X \leq 30$.
(d) $0 \leq X \leq 10$ or $20 \leq X \leq 30$.

46. A national poll found that 45% viewed the media as too liberal, 40% just right, 10% as too conservative, while 5% had no opinion. A local poll of 500 found 250 viewed the media as too liberal, 175 as just right, 50 as too conservative, while 25 had no opinion. Test at $\alpha = 0.10$ that the local opinion of the media differs from the national opinion of the media. Which of the following statements are true?

    (a) The expected numbers are 225, 200, 50, and 25.

    (b) The computed test statistic is 5.90278.

    (c) The chi-squared test statistic has 4 degrees of freedom.

    (d) The p-value is 0.1164.

47. The amount of time patients spent with the doctor in about 880 million office visits in 2001 according to the National Center for Health Statistics was:
    - 1–10 minutes, 22.9%
    - 11–15 minutes, 36.0%
    - 16–30 minutes, 30.3%
    - 31 minutes or more, 6.5%
    - No time with doctor, 4.3%

    One thousand patient record visits were selected from Ace Health Care systems to see if their time distribution differed from the national figures:
    - 1–10 minutes, 250
    - 11–15 minutes, 375
    - 16–30 minutes, 285
    - 31 minutes or more, 60
    - No time with doctor, 30

    The above data are used to test whether the distribution is the same as the national distribution at $\alpha = 0.05$. Which of the following statements are true?

    (a) The critical value is 7.77944.

    (b) The expected numbers are 229, 360, 303, 65, and 43.

    (c) The computed test statistic is 7.93492.

    (d) The p-value is 0.09399.

48. A study concerning the relationship between male/female supervisory structure and the level of employee's job satisfaction was performed, with the results shown in Table 10. The Minitab analyses of the data are shown below.

Table 10 Male/female supervisory structure and employee's job satisfaction.

Level of satisfaction	Boss/Employee			
	Female/Male	Female/Female	Male/Male	Male/Female
Satisfied	33	20	35	35
Neutral	25	35	28	25
Dissatisfied	17	45	25	20

Chi-Square Test: Female/Male, Female/Female, Male/Male, Male/Female

```
Expected counts are printed below observed counts
Chi-Square contributions are printed below expected counts
       Female/Male Female/Female Male/Male Male/Female Total
1        33            20            35         35        123
        26.90         35.86         31.56      28.69
         1.386         7.015         0.376      1.389

2        25            35            28         25        113
        24.71         32.94         28.99      26.36
         0.003         0.128         0.034      0.070

3        17            45            25         20        107
        23.40         31.20         27.45      24.96
         1.749         6.109         0.219      0.984

Total 75            100            88         80        343
Chi-Sq = 19.461, DF = 6, P-Value = 0.003
```

Which of the following statements are true?
(a) The table has 16 cells.
(b) The area under the chi-squared curve having 6 degrees of freedom to the right of 19.461 is 0.003.
(c) This test of hypothesis is always a two-tailed test.
(d) For $\alpha = 0.05$, do not reject the null hypothesis that male/female supervisory structure is independent of the level of employee's job satisfaction.

49. A dental research project was conducted to determine whether the tooth-bleaching method being used was independent of the gender of the participant. *Dentists* cover the gums and apply peroxide to the teeth; cost: $250–$500. *Plastic trays* are flexible, thin plastic

Table 11 Relationship between tooth-bleaching method and gender of patient.

		Tooth-bleaching method	
Gender	Dentist	Plastic tray	Strips, wands, and toothpaste
Male	15	50	35
Female	45	25	30

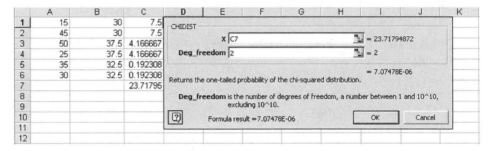

Fig. 8.

molds that fit around the teeth; cost: $200–$350. *Strips* fit across the teeth like address labels; cost: $25–$35 at stores, $40–$50 from dentists. Wands and toothpastes whiten teeth for a few dollars. The data is given in Table 11. The Excel analysis is shown in Fig. 8. Which of the following statements are true?

(a) The computed test statistic is 23.71795.

(b) The computed p-value is 0.00000707477.

(c) The terms in column C are the terms in the sum $\chi^2 = \sum (f_{ij} - e_{ij})^2 / e_{ij}$.

(d) The tooth-bleaching method selected is dependent on the gender of the participant at $\alpha = 0.05$.

50. A correlation study looked at the connection between the hours spent watching TV by teenagers and their weight. Table 12 gives the hours per week spent watching TV and the following coded weight measurements: 0 if the participant is not overweight, 1 if 10 pounds or less overweight, 2 if between 10 and 20 pounds overweight, 3 if between 20 and 30 pounds overweight, and 4 if more than 30 pounds overweight.

Table 12 Relationship between weight and TV watching time for teenagers.

Hours of TV per week, x	Coded weight, y	Rank of x	Rank of y
25	3	11	10
15	3	5.5	10
20	4	8.5	13.5
5	2	1.5	5.5
30	4	13.5	13.5
15	2	5.5	5.5
20	2	8.5	5.5
15	3	5.5	10
25	1	11	2.5
30	3	13.5	10
25	2	11	5.5
5	0	1.5	1
10	1	3	2.5
15	3	5.5	10

Which of the following statements are true?

(a) The Pearson correlation coefficient finds the correlation coefficient between x and y (columns 1 and 2).

(b) The Spearman rank correlation coefficient finds the correlation coefficient between the ranks of x and y (columns 3 and 4).

(c) The Spearman rank correlation coefficient is always between -1 and 1.

(d) The Pearson correlation coefficient is 0.473 and the Spearman rank correlation coefficient is 0.519.

Answers to Final Exam I

1. b
2. c
3. d
4. c
5. a
6. d
7. a
8. a, b, c
9. a, d
10. a, c
11. a, c, d
12. b, c
13. b, c
14. a, c
15. c
16. a, b
17. b, c, d
18. b, d
19. c, d
20. a, b, c
21. a, c, d
22. a, b, d
23. c
24. a
25. a, c
26. a, c
27. a, b, c, d
28. a, b, c
29. a, b, c, d
30. b, c
31. d
32. b, c
33. a, b, c
34. a, b, c
35. c, d
36. d
37. b, c, d
38. b, c

39. d
40. a, b, c, d
41. a, b, c, d
42. a, b, c, d
43. a, b, c, d
44. a
45. a
46. a, b, d
47. b, c, d
48. b
49. a, b, c, d
50. a, b, c

Final Exam II

1. Give the Excel command and the Minitab command to evaluate the alpha level for the rejection region shown in Fig. 9.

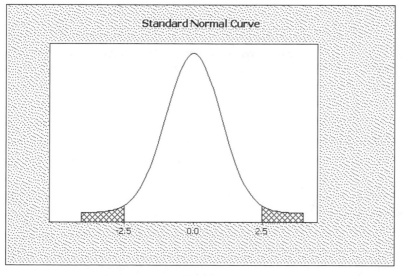

Fig. 9.

2. Give the Excel command and the Minitab command to evaluate the alpha level for the rejection region shown in Fig. 10.

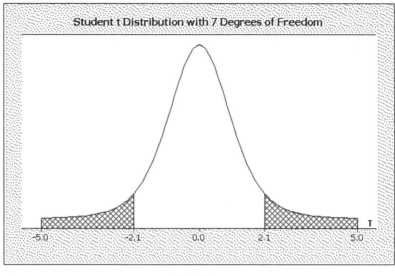

Fig. 10.

3. Give the Excel command and the Minitab command to evaluate the alpha level for the rejection region shown in Fig. 11.

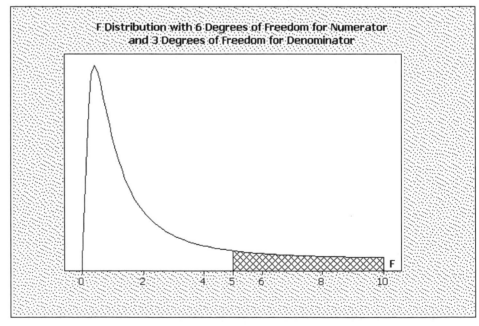

Fig. 11.

4. Give the Excel command and the Minitab command to evaluate the alpha level for the rejection region shown in Fig. 12.

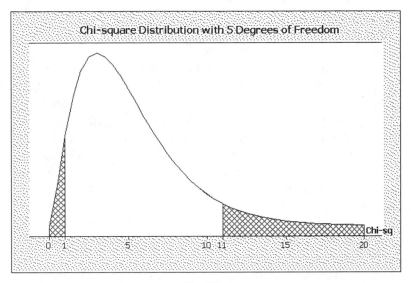

Fig. 12.

5. If α is given and the research hypothesis is known (that is, one- or two-tailed), then the inverse function may be used to find the rejection region. The Excel inverse functions are as follows: =NORMSINV, =TINV, =FINV, and =CHIINV. In Minitab, the dialog boxes for Normal, t, F, and Chi-sq also contain the inverse functions. Suppose the null hypothesis is H_0: $\mu = \mu_0$ and the alternative hypothesis is H_a: $\mu \neq \mu_0$ and $\alpha = 0.05$ and the sample is large. Find the rejection region.

6. Give the Excel command to find α if the rejection region is $Z \leq -2.33$.

7. Give the Minitab pull-down used to perform a single sample test for a population mean for a large sample.

8. Give the Minitab command to find α if the rejection region is $Z \leq -2.33$.

9. Give the Minitab pull-down used to perform a single sample test for a population mean for a small sample.

10. Give the Excel command to find α if the rejection region is $Z \leq -2.00$ or $Z \geq 2.00$.

11. Give the Minitab pull-down used to perform a single sample test for a population proportion for a large sample.

12. Give the Minitab commands and the value of the α level if the rejection region is $Z \leq -2.00$ *or* $Z \geq 2.00$.

13. Use Excel to find the rejection region if $\alpha = 0.15$ and the research hypothesis is $\mu \neq \mu_0$ for a large sample test.

14. Give the Excel command to find α if the rejection region is $T \leq -2.19$ and T is a student t variable with 11 degrees of freedom.

15. Give the Excel command to find α if the rejection region is $T \leq -3.00$ *or* $T \geq 3.00$ and T is a student t variable with 7 degrees of freedom.

16. Use Excel to find the rejection region if $\alpha = 0.1$ and the research hypothesis is $\mu \neq \mu_0$ for a small sample test with $n = 13$.

17. Use Excel to find the rejection region if $\alpha = 0.1$ and the research hypothesis is $\mu < \mu_0$ for a small sample test with $n = 13$.

18. Give the Minitab command and the value of the α level if the rejection region is $T \leq -2.19$ and T is a student t variable with 11 degrees of freedom.

19. Give the Minitab command and the value of the α level if the rejection region is $T \leq -3.00$ *or* $T \geq 3.00$ and T is a student t variable with 7 degrees of freedom.

20. Use Minitab to find the rejection region if $\alpha = 0.1$ and the research hypothesis is $\mu < \mu_0$ for a small sample test with $n = 13$.

21. Use Minitab to find the rejection region if $\alpha = 0.1$ and the research hypothesis is $\mu \neq \mu_0$ for a small sample test with $n = 13$.

22. Suppose you are testing that the mean amount spent on dentistry per year for adults is \$750 versus it is not (H_0: $\mu = \$750$ versus H_a: $\mu \neq \$750$). A sample of size 400 is taken and the test statistic is equal to $Z = 3.45$. Use Excel to find the p-value for this test statistic.

23. Suppose you are testing that the mean amount spent on dentistry per year for adults is \$750 versus it is not (H_0: $\mu = \$750$ versus H_a: $\mu \neq \$750$). A sample of size 400 is taken and the test statistic is equal to $Z = 3.45$. Use Minitab to find the p-value for this test statistic.

24. The variance of the amount of fill in an automatic filling machine is hypothesized to be less than 1 ounce-squared (H_0: $\sigma^2 = 1$ versus H_a: $\sigma^2 < 1$). A sample of 25 containers filled by the machine is found to have $S^2 = 0.55$. The computed test statistic is $(n-1)S^2/\sigma_0^2 = 13.2$. Find the p-value, using Excel.

25. Find the p-value in problem 24 using Minitab.

26. The pH of Metro city drinking water is of interest. The city has a target value of 8.0. A sample of size 15 is selected and it is found that the sample mean equals 8.12 and the standard deviation

is 0.15. The computed test statistic for testing H_0: $\mu = 8.0$ versus H_a: $\mu \neq 8.0$ is 3.10. Find the p-value for this test, using Excel.

27. Find the p-value in problem 26 using Minitab.

28. Twenty patients were randomly divided into two groups of ten each. One group was placed on diet 1 and the other on diet 2. The weight losses and variances for each group were recorded. Using the two-sample t-test assuming equal variances, the hypothesis H_0: $\mu_1 - \mu_2 = 0$ versus H_a: $\mu_1 - \mu_2 \neq 0$ was tested. If the computed test statistic was equal to 1.88, find the p-value for the test, using Excel.

29. Find the p-value in problem 28 using Minitab.

30. An experiment was designed to compare two diets with respect to the variation in weight losses in the two groups. It was determined that the weight losses were normally distributed in both groups. The summary statistics for the two diets were as follows: diet 1: $n_1 = 13$, $S_1 = 14.3$; diet 2: $n_2 = 16$, $S_2 = 8.9$. The hypotheses were H_0: $\sigma_1 = \sigma_2$ and H_a: $\sigma_1 > \sigma_2$. The computed test statistic was $F = 2.582$. Use Minitab to compute the p-value.

31. Use the paste function, FDIST, of Excel to find the p-value in problem 30.

32. A survey of 500 men and 500 women found that 44% of men and 38% of women shop online for Christmas gifts. This survey was used to test H_0: $p_1 - p_2 = 0$ versus H_a: $p_1 - p_2 \neq 0$. Find the value of the test statistic and the p-value for this test, using the function NORMSDIST of Excel.

33. Four treatments were compared with respect to their ability to shorten the duration of a cold. Eighty individuals were randomly divided into four groups of 20 each and the four treatments were compared with respect to the time required for a cold to run its course. The F value for the hypothesis of no difference in the four means was 2.68. Use Excel to find the p-value for testing the null hypothesis H_0: $\mu_1 = \mu_2 = \mu_3 = \mu_4$.

34. Use Minitab to find the p-value in problem 33.

35. A research study looked at the connection between obesity in children and in their parents. Table 13 gives the number of pounds

Table 13

y	35	15	40	25	25	50	25	37	49	35	40	40	15	40	10
x	20	25	30	25	20	40	15	32	37	25	35	40	15	50	15

above a healthy weight for pairs each comprising a father and his oldest child. x is the number of excess pounds for the father and y is the number of excess pounds for the oldest child. Use Excel and Minitab to find the equation of the linear regression equation that connects y to x.

36. In problem 35, use Minitab to give the 99% prediction interval and 99% confidence interval for the amount overweight of the oldest child if the father is 30 pounds overweight.

37. In problem 35, give a point estimate for β_1, the slope of the regression line. Also, find a 95% confidence interval for β_1. Remember that $\hat{\beta}_1$ is the additional amount overweight that a child would be for each additional pound that the child's father is overweight.

38. Table 14 gives the prices of homes in Cape Sanibel, Florida. Also given is the following information: the number of bathrooms, the number of bedrooms, the square footage, and whether or not the home is located with a canal in the backyard that connects the home to the Gulf ($0 =$ no, $1 =$ yes). Find the least-squares prediction equation.

39. From the data in problem 38, find the prediction interval and confidence interval for the price of a home that has three bedrooms, two baths, 2000 square feet, and a canal connection.

40. The following output is for a stepwise regression performed on the data in problem 38.

Stepwise Regression: price versus bedroom, bath, footage, connection
```
Forward selection. Alpha-to-Enter: 0.15
Response is price on 4 predictors, with N = 15
```

Step	1	2
Constant	−254.2	−402.8
footage	0.298	0.181
T-Value	6.31	3.29
P-Value	0.000	0.006
bedrooms		129
T-Value		2.92
P-Value		0.013
S	95.9	76.3
R-Sq	75.39	85.59
R-Sq(adj)	73.50	83.19
Mallows C-p	10.7	3.7

Table 14 Information on Florida homes.

Price, $(thousands)	# Bedrooms	# Baths	Square footage	Canal connection
200	2	2	1500	0
300	3	2	2000	0
350	3	3	2000	0
400	3	3	2500	1
500	4	3	2500	1
240	3	3	2500	0
475	4	3	2000	1
325	3	3	1750	1
175	2	2	1500	0
600	4	4	2500	1
325	3	2	2000	0
750	4	4	3000	1
800	4	4	3500	1
325	3	2	2000	0
475	3	3	2500	1

Give the best straight line fit to the data in two dimensions and the best plane fit to the data in three dimensions.

41. Refer to problem 40. What percent of the variation in prices is accounted for by the straight-line model? What percent of the variation in prices is accounted for by the model that represents a plane?

42. A four-sided object with the numbers 1, 2, 3, and 4 painted on the sides is tossed 20 times. If the object is balanced, each side rests on

the ground with probability 0.25 on each toss. Let $X =$ the number of times side 4 rests on the ground in the 20 tosses.

Suppose we wish to test H_0: $p = 0.25$ versus H_a: $p \neq 0.25$ and the rejection region is $X = 0$, 1, 9, 10, 11, 12, 13, 14, 15, 16, 17, 18, 19, 20. Find the value of α.

43. It is reported that the number of computers that are in homes in a particular section of the country follows the distribution: 0, 5%; 1, 35%; 2, 30%; 3 or more 30%. A survey of 1000 is conducted and a goodness-of-fit test is performed. The computed test statistic equals 7.45. Find the *p*-value for this test, using Excel.

44. In problem 43, find the *p*-value using Minitab.

45. A survey of Internet users was performed. One question asked for the number of orders placed on the Internet during the past year and the other asked for the number of spam messages received weekly. The results are given in a table similar to Table 15. Suppose that none of the expected cell frequencies are less than 5 and that the value of the following $\chi^2 = \sum (f_{ij} - e_{ij})^2 / e_{ij}$ over all 20 cells was equal to 13.45. The null hypothesis is that the number of spam messages received weekly is independent of the number of Internet orders placed during the past year. Use Excel to find the *p*-value.

Table 15 Relation between level of Internet shopping and level of spam received.

Spam/week	Number of Internet orders during the past year				
	0–5	6–10	11–15	16–20	Over 20
0–25					
26–50					
50–75					
Over 75					

46. Refer to problem 45. Use Minitab to find the answer to the problem.

47. Twenty-five people are selected and asked to try two after-shave lotions. They are asked to try brand 1 on one side and brand 2 on

the other, then to say which brand they prefer. They cannot like the two brands equally. The null hypothesis is H_0: there is no difference in the brands, and the research hypothesis is H_a: there is a difference in the brands. They are to analyze the data using the nonparametric sign test. Let $X =$ the number who prefer brand 1 over brand 2. Give the rejection region if $\alpha = 0.10$ and they are not to go over this value but to get as close as possible without going over. The rejection region is to be divided equally on both sides of the mean.

48. Twenty-five patients who needed to have their appendices removed had traditional operations while thirty who needed to have their appendices removed had laparoscopic appendectomies. The number of days of hospital stay was recorded for each of the 55 patients. Suppose every patient in the laparoscopic group had shorter hospital stays than the patients in the traditional group. The Wilcoxon rank sum test was used to compare the two groups. Calculate the normal approximation Z statistic corresponding to the laparoscopic group.

49. A paired design was used to test whether husbands and wives spend the same amount of time on the Internet on the average. Thirty husband/wife couples were asked to keep a diary of time spent on the Internet weekly. The data is shown in Table 16.

Table 16   Weekly time spent on the Internet by 30 husband/wife pairs.

Husband	Wife	Difference	Husband	Wife	Difference
20	9	11	14	15	−1
17	10	7	16	7	9
17	9	8	18	5	13
13	9	4	18	9	9
11	7	4	19	8	11
12	9	3	12	8	4
12	10	2	18	11	7
21	11	10	15	12	3

(Continued)

Table 16 Continued.

Husband	Wife	Difference	Husband	Wife	Difference
15	4	11	15	11	4
15	9	6	17	10	7
15	9	6	13	6	7
10	10	0	11	14	−3
15	13	2	12	7	−4
13	15	−2	18	6	12
23	9	14	11	10	1

The paired t-test is used to test H_0: $\mu_D = 0$ versus the research hypothesis H_a: $\mu_D \neq 0$. Compute the test statistic and give the Excel command that computes the p-value. Give the value of the computed p-value.

50. Refer to problem 49. Suppose the nonparametric procedure called the Wilcoxon signed rank test is used to analyze the data. Give the Minitab Wilcoxon signed rank test analysis.

Answers to Final Exam II

1. Excel command $= \text{NORMSDIST}(-2.5) + (1\text{-NORMSDIST}(2.5))$
   0.012419

   The Minitab pull-down **Calc** $\Rightarrow$ **Probability Distributions** $\Rightarrow$ **Normal Distribution** gives the dialog box shown in Fig. 13, which is filled as shown. The following output is given.

   Cumulative Distribution Function
   ```
   Normal with mean = 0 and standard deviation = 1
   x              P(X <= x)
   -2.5           0.0062097
   ```

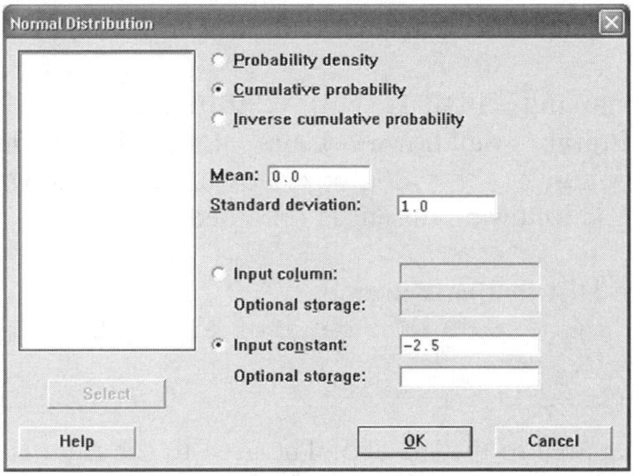

Fig. 13.

Because the curve is symmetrical, we double 0.0062097 to obtain 0.0124194.

2. Excel command = TDIST(2.1,7,2)   0.073871
    The Minitab pull-down **Calc ⇒ Probability Distributions ⇒ t Distribution** gives the dialog box shown in Fig. 14, which is filled as shown. The following output is obtained.

Cumulative Distribution Function
```
Student's t distribution with 7 DF
x              P(X <= x)
−2.1           0.0369356
```

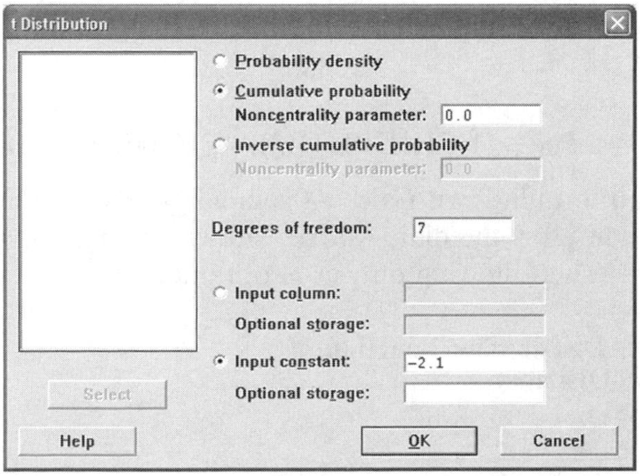

Fig. 14.

Because the curve is symmetrical, we double 0.0369356 to obtain 0.073871.

3. Excel command = FDIST(5,6,3)   0.107262

    The Minitab pull-down **Calc ⇒ Probability   Distributions ⇒ F Distribution** gives the dialog box shown in Fig. 15, which is filled as shown. The following output is obtained.

Cumulative Distribution Function
```
F distribution with 6 DF in numerator and 3 DF in denominator
x             P(X <= x)
5             0.892738
```

This is the area to the left of 5. The area to the right is $1 - 0.892738 = 0.107262$.

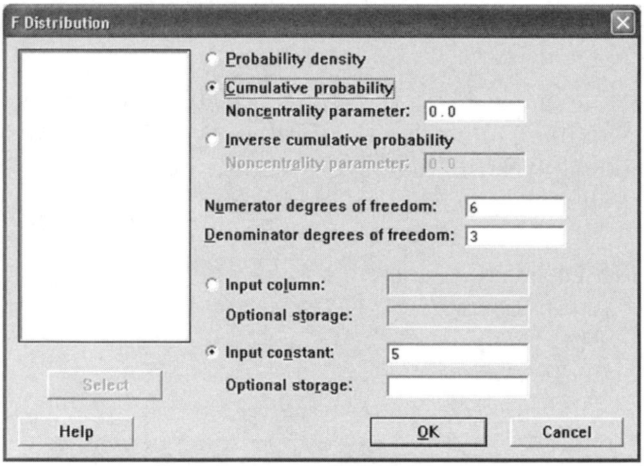

Fig. 15.

4. Excel command = (1-CHIDIST(1,5))+CHIDIST(11,5) 0.088814

    The Minitab pull-down **Calc ⇒ Probability Distributions ⇒ Chisquare Distribution** gives the dialog boxes shown in Fig. 16, which are filled as shown. The following output is obtained.

Cumulative Distribution Function
```
Chi-Square with 5 DF
x             P(X <= x)
1             0.0374342
```

and

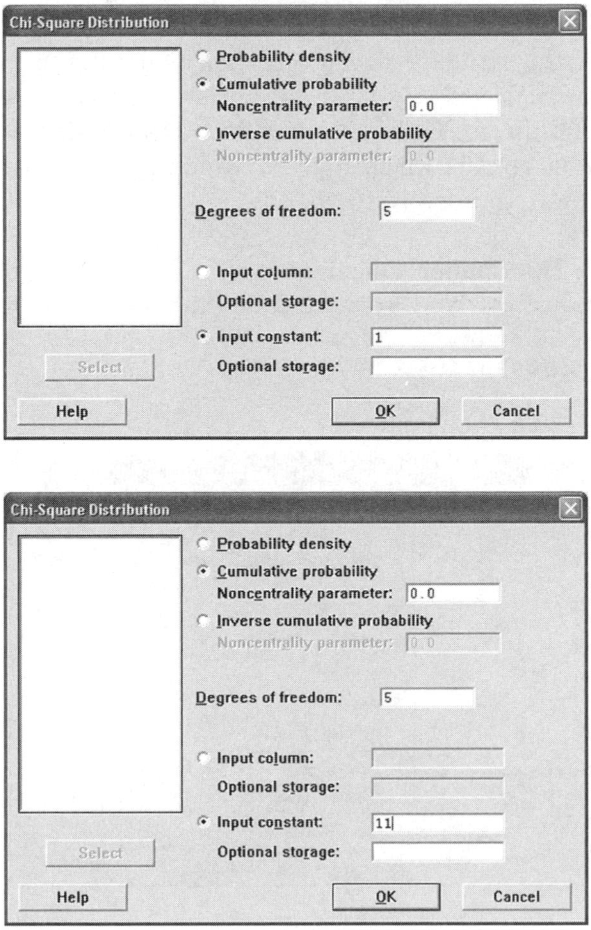

Fig. 16.

Cumulative Distribution Function

```
Chi-Square with 5 DF
x              P(X <= x)
11             0.948620
```

The area to the right of 11 is $1 - 0.948620 = 0.05138$. This is added to 0.0374342 to obtain 0.0888142.

5. =NORMSINV(0.025)   −1.95996 is the lower critical value.
   =NORMSINV(0.975)   1.95996 is the upper critical value.
6. =normdist(−2.33) given in any cell   0.009903

7. **Stat ⇒ Basic Statistics ⇒ 1 Sample Z** (Note that you need an estimate of the standard deviation when you enter the dialog box for this routine).

8. The pull-down **Calc ⇒ Probability Distribution ⇒ Normal** gives the dialog box shown in Fig. 17, which is filled out as shown. The following output is produced.

Cumulative Distribution Function

```
Normal with mean = 0 and standard deviation = 1
x              P(X <= x)
-2.33          0.0099031
```

Fig. 17.

9. **Stat ⇒ Basic Statistics ⇒ 1-sample t**

10. =normsdist(−2) + (1 − normsdist(2))
(this is one solution)   0.0455

11. **Stat ⇒ Basic Statistics ⇒ 1-proportion**

12. Use the pull-down **Calc ⇒ Probability Distribution ⇒ Normal** to find $P(Z \leq -2.00)$. Use the same pull-down to find $P(Z \leq 2.00)$ and then find $P(Z > 2.00) = 1 - P(Z \leq 2.00)$ using a calculator. Now add the two.   0.0455

13. =normsinv(0.075) will give the left-hand critical value, −1.43953. By symmetry of the standard normal curve, the right-hand value is 1.43953. The rejection region is $|Z| > 1.43953$.

14. =tdist(2.19,11,1)   0.0255   The function gives the alpha level for critical value 2.19 and degrees of freedom 11. The 1 means the area in one tail. By symmetry, the area to the left of −2.19 is the same as the area to the right of 2.19. The tdist function is a little more complicated than the normsdist function but is still easy to use once you see how it is structured.

15. =tdist(3.00,7,2)   0.019942   The 2 in tdist gives the area to the left of –3.00 plus the area to the right of 3.00.

16. =TINV(0.1,12)   1.782287   The area to the left of −1.782287 plus the area to the right of 1.782287 is 0.1. That is, the area to the left of −1.782287 is 0.05 and the area to the right of 1.782287 is 0.05.

17. =TINV(0.2,12)   1.356218   The area to the left of −1.356218 plus the area to the right of 1.356218 is 0.2. The area to the left of −1.356218 is 0.1 and the area to the right of 1.356218 is 0.1. Because this is a lower-tail test at $\alpha = 0.1$, the rejection region is $T < -1.356218$.

18. The pull-down **Calc $\Rightarrow$ Probability Distribution $\Rightarrow$ t** gives the t Distribution dialog box shown in Fig. 18, which is filled out as shown. The output is as follows.

```
Student's t distribution with 11 DF
x              P(X <= x)
−2.19          0.0254841
```

$\alpha = 0.0254841.$

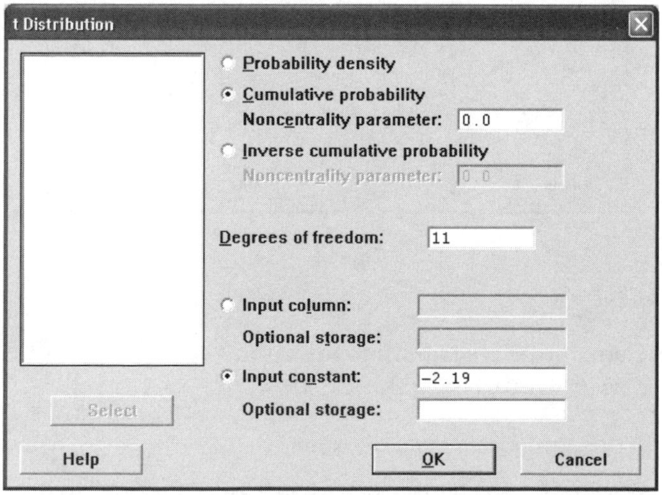

Fig. 18.

19. The pull-down **Calc ⇒ Probability Distribution ⇒ t** gives the t Distribution dialog box as in problem 18. By entering 7 for degrees of freedom and −3 for the input constant, we see that the area to the left of −3 is 0.0099711.

```
Student's t distribution with 7 DF

x                  P(X <= x)
-3                 0.0099711
```

Because the t curve is symmetrical the area to the right of 3 is the same as the area to the left of −3, we see that $\alpha = 2(0.0099711) = 0.0199422$.

20. The dialog box is shown in Fig. 19. Notice that Inverse cumulative probability is selected and the area in the left tail is given as the input constant.

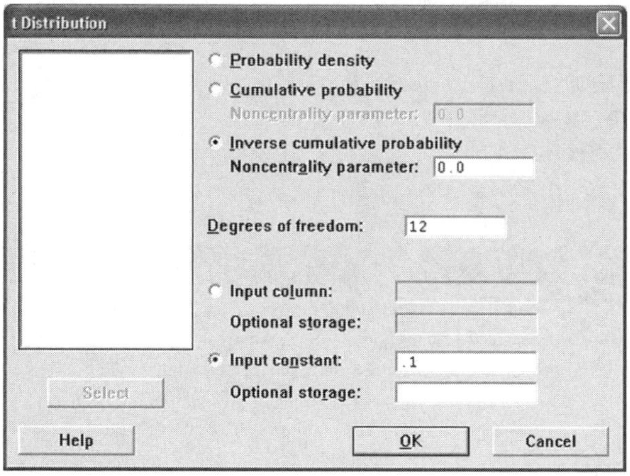

Fig. 19.

Inverse Cumulative Distribution Function
```
Student's t distribution with 12 DF
P(X <= x)     x
0.1           -1.35622
```

The area under the curve to the left of −1.35622 is 0.1. The critical value is −1.35622 and the rejection region is $t < -1.35622$.

21. Because this is a two-tailed test, the alpha is split and half is assigned to each tail. As in problem 20, the degrees of freedom is entered as 12 and 0.05 is given as the input constant.

Inverse Cumulative Distribution Function
```
Student's t distribution with 12 DF
P(X <= x)    x
0.05         -1.78229
```

> There is 0.05 area to the left of -1.78229 and 0.05 area to the right of 1.78229. The rejection region is $|T| > 1.78229$.

22. p-value = twice the area to the right of 3.45. $= 2*(1-$ NORMSDIST(3.45)) $\quad$ 0.000561.

23. Find the area to the left of -3.45 and double it.

Cumulative Distribution Function
```
Normal with mean = 0 and standard deviation = 1
x            P(X <= x)
-3.45        0.0002803
```

p-value $= 2(0.0002803) = 0.0005606$.

24. =CHIDIST(13.2,24) $\quad$ 0.962709 gives the area to the right of 13.2. Since this is a lower-tail test the p-value is $1 - 0.962709 = 0.037291$.

25. The pull-down **Calc** $\Rightarrow$ **Probability Distribution** $\Rightarrow$ **Chi-Square** gives the dialog box shown in Fig. 20. It is filled out as shown. The output produced by this dialog box is

Cumulative Distribution Function
```
Chi-Square with 24 DF
x            P(X <= x)
13.2         0.0372908
```

The p-value corresponding to the test statistic value 13.2 is 0.0372908.

26. =TDIST(3.1,14,2) $\quad$ p-value $= 0.007832$.

27. **Cumulative Distribution Function**
```
Student's t distribution with 14 DF
x            P(X <= x)
-3.1         0.0039162
```

The p-value is $0.0039162 + 0.0039162 = 0.007832$.

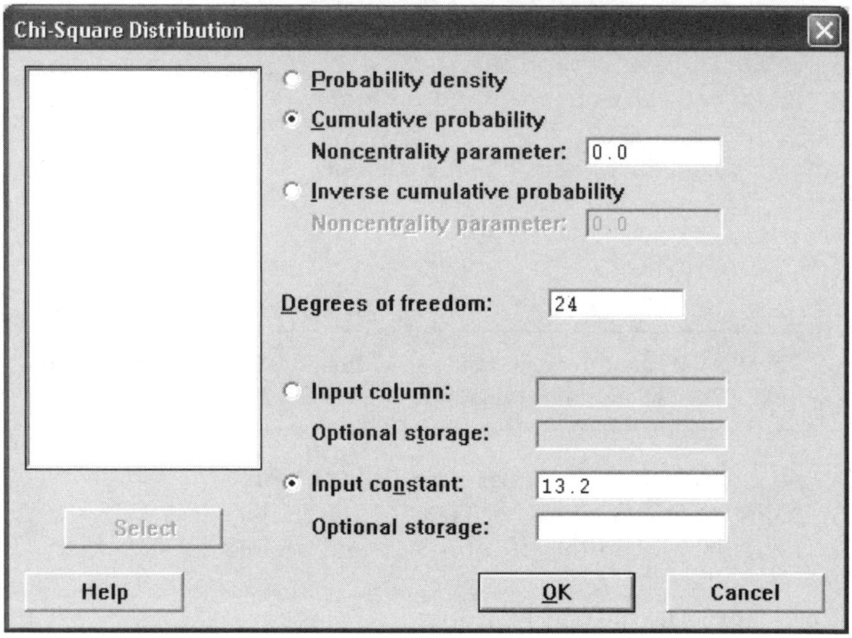

Chi-Square Distribution   ☒

- ○ Probability density
- ⊙ Cumulative probability
  Noncentrality parameter: [0.0]
- ○ Inverse cumulative probability
  Noncentrality parameter: [0.0]

Degrees of freedom: [24]

- ○ Input column: [   ]
  Optional storage: [   ]

- ⊙ Input constant: [13.2]
  Optional storage: [   ]

[Select]

[Help]   [OK]   [Cancel]

Fig. 20.

28. =TDIST(1.88,18,2)   0.076393.

29. **Cumulative Distribution Function**
     Student's t distribution with 18 DF.

```
x                P(X <= x)
-1.88            0.0381966
```

The p-value is $0.0381966 + 0.0381966 = 0.076393$.

30. The pull-down **Calc** $\Rightarrow$ **Probability Distributions** $\Rightarrow$ **F** gives the dialog box shown in Fig. 21. The box is filled out as shown, and the following output is produced.

Cumulative Distribution Function

```
F distribution with 12 DF in numerator and 15 DF in denominator
x                P(X <= x)
2.582            0.957206
```

Since this is an upper-tailed test, the p-value $= 1 - 0.957206 = 0.042794$.

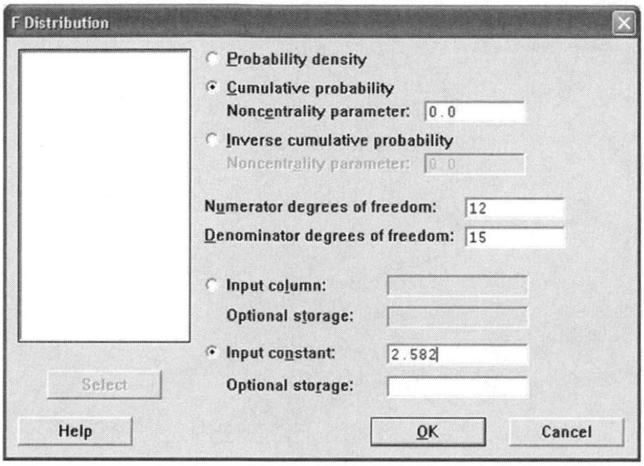

Fig. 21.

31. FDIST gives the dialog box shown in Fig. 22.

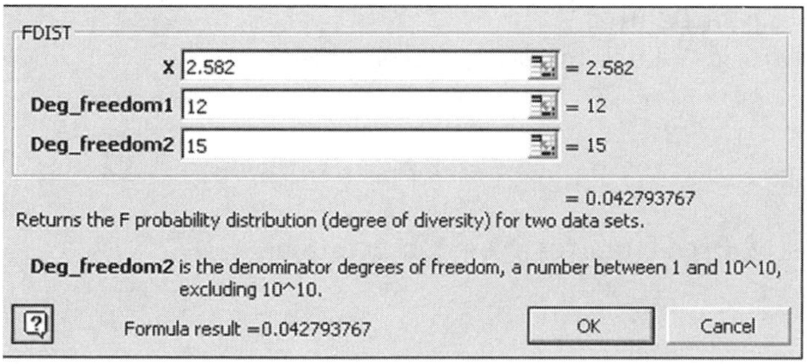

Fig. 22.

The formula result, 0.042794, is the area to the right of 2.582 and is the *p*-value.

32. $Z = 1.93$ $= 2*(1\text{-NORMSDIST}(1.93))$ 0.053607

33. The *F* value has 3 and 76 degrees of freedom and its value is 2.68. The *p*-value is the area under the *F* curve to the right of 2.68.
     $= \text{FDIST}(2.68,3,76)$ *p*-value $= 0.052822$

34. The pull-down **Calc $\Rightarrow$ Probability Distributions $\Rightarrow$ F** gives Fig. 23, which is filled out as shown. The following output is generated.

Cumulative Distribution Function

```
F distribution with 3 DF in numerator and 76 DF in denominator
x               P(X <= x)
2.68            0.947178
```

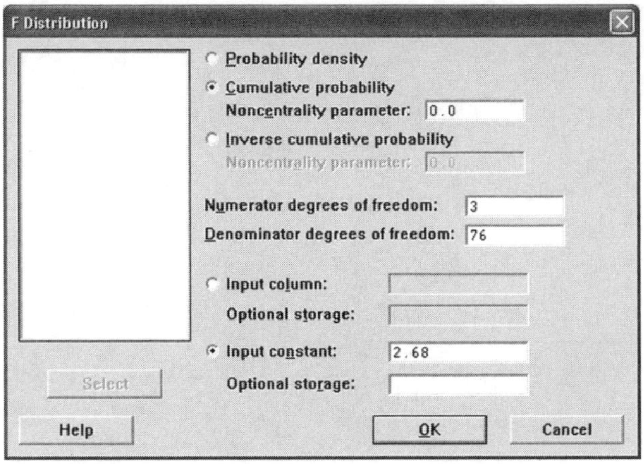

Fig. 23.

The area to the right of 2.68 is $1 - 0.947178 = 0.052822$.

35. $y = 6.61 + 0.901x$.

36.   **Predicted Values for New Observations**

```
New
Obs    Fit      SE Fit   99% CI          99% PI
1      33.63    2.10     (27.30, 39.96)  (8.65, 58.61)
```

Values of Predictors for New Observations

```
New
Obs    x
1      30.0
```

The prediction interval for the amount overweight for a child whose father is 30 pounds overweight is between 8.65 pounds and 58.61 pounds. The confidence interval for the mean amount overweight for all children whose fathers are 30 pounds overweight is between 27.30 and 39.96 pounds overweight.

37.   0.9 is the point estimate of β_1. The 95% confidence interval for β_1 is $0.9 \pm 2.160(0.202)$; i.e., the 95% interval extends from 0.464 to 1.336.

38.   price $= -327 + 84.7$ bedrooms $+ 38.7$ baths $+ 0.146$ footage $+ 59.7$ connect

39.   **Predicted Values for New Observations**

```
New
Obs    Fit      SE Fit   95% CI          95% PI
1      356.8    65.1     (211.9, 501.8)  (136.9, 576.7)
```

Values of Predictors for New Observations

```
New
Obs    bedrooms        baths          footage        connect
1      3.00            2.00           2000           1.00
```

The predicted price for a home having three bedrooms, two baths, 2000 square feet, and a canal connection is $356,800. The 95% prediction interval for such a home is $136,900 to $576,700. The 95% confidence interval for the mean price of all homes having three bedrooms, two baths, 2000 square feet, and a canal connection is from $211,900 to $501,800.

40. Price $= -254.2 + 0.298$ footage.
    Price $= -402.8 + 0.181$ footage $+ 129$ bedrooms.

41. Accept either 75.39% or 73.50% for the straight-line model.
    Accept either 85.59% or 83.19% for the planar model.

42. =BINOMDIST(1,20,0.25,1) + (1-BINOMDIST(8,20,0.25,1))
    0.065238

43. =CHIDIST(7.45,3)   0.058857
    This command gives the area under the chi-squared curve having 3 degrees of freedom from 7.45 to the right. The p-value $= 0.058857$.

44. The pull-down **Calc ⇒ Probability Distributions ⇒ Chi-Square** gives the dialog box shown in Fig. 24. Fill in the dialog box as shown. The following output is obtained.

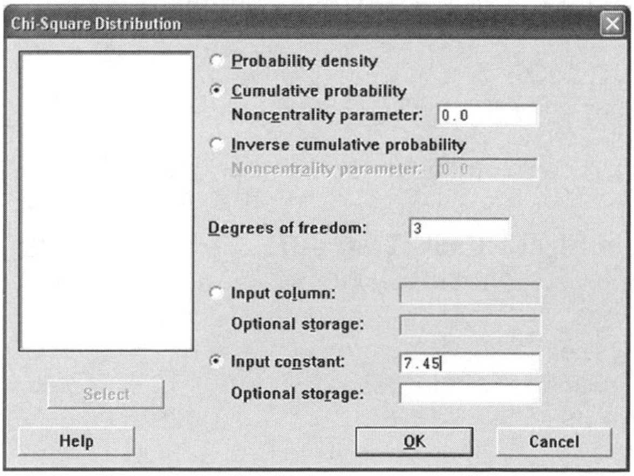

Fig. 24.

Cumulative Distribution Function

```
Chi-Square with 3 DF
x        P(X <= x)
7.45     0.941143
```

This gives areas from 7.45 to the left. We need the area to the right because this is an upper-tailed test. The p-value is $1 - 0.941143 = 0.058857$.

45. =CHIDIST(13.45,12)   0.337198

This command gives the area under the chi-squared curve having 12 degrees of freedom from 13.45 to the right. The p-value $= 0.337198$.

46. **Cumulative Distribution Function**

```
Chi-Square with 12 DF
x        P(X <= x)
13.45    0.662802
```

Since this is an upper-tailed test, the p-value $= 1 - 0.662802 = 0.337198$.

47. The rejection region is $X = 0$ through 7 or $X = 13$ through 20, and $\alpha = 0.043285$.

48. $T_1 = 465$

$$\mu_T = \frac{n_1(n_1 + n_2 + 1)}{2} = \frac{30(30 + 25 + 1)}{2} = 840$$

$$\sigma_T = \sqrt{\frac{n_1 n_2 (n_1 + n_2 + 1)}{12}} = \sqrt{\frac{30(25)(56)}{12}} = 59.16$$

$$Z = \frac{465 - 840}{59.16} = -6.34$$

49. Computed test statistic $= 7.06$.   =TDIST(7.06,29,2)   9.1287E-0850.

50. **Wilcoxon Signed Rank Test: diff**

```
Test of median = 0.000000 versus median not = 0.000000
```

	N	N for Test	Wilcoxon Statistic	P	Estimated Median
diff	30	29	411.5	0.000	5.500

Solutions to Chapter Exercises

Introduction

1. (a) 0.0301   (b) 0.0071   (c) 0.0340.
2. (a) $Z > 1.44$ (b) $Z < -1.175$ (c) $|Z| > 1.37$.
3. (a) p-value $= 0.0307$ (b) p-value $= 0.0071$ (c) p-value $= 0.0105$.
4. (a) Reject the null hypothesis.   (b) Unable to reject the null hypothesis.   (c) Reject the null hypothesis.
5. (a) H_0: $\mu = 5$ H_a: $\mu \neq 5$
    (b) Do not reject because 5 is included in the 95% confidence interval for μ.
    (c) Do not reject because the computed test statistic, $Z = -0.17$, is not in the rejection region.
    (d) Do not reject because the p-value $> \alpha$.
6. (a) 0.025   (b) 0.005   (c) 0.002

7. (a) $t > 1.761$ (b) $t < -1.345$ (c) $|t| > 2.977$
8. (a) p-value $= 0.05$ (b) p-value $= 0.025$ (c) p-value $= 0.02$
9. (a) H_0: $\mu = 105$ versus H_a: $\mu \neq 105$
   (b) Do not reject the null hypothesis because 105 is contained in the 95% confidence interval.
   (c) The rejection region is $|t| > 2.093$ and -2.02 is not in the rejection region. Do not reject the null hypothesis.
   (d) The p-value is 0.058 and it is greater than 0.05. Do not reject the null hypothesis.
10. (a) H_0: $p = 5\%$ versus H_a: $p \neq 5\%$
    (b) Do not reject the null hypothesis because 5% is in the 95% confidence interval.
    (c) The rejection region is $|Z| > 1.96$ and 1.69 is not in the rejection region. Do not reject the null hypothesis.
    (d) The p-value is 0.092 and it is greater than 0.05. Do not reject the null hypothesis.
    (e) Yes, $np_0 = 7.5 > 5$ and $nq_0 = 142.5 > 5$.
11. ($160.57, $426.17).
12. $Z = -1.75$, p-value $=$ NORMSDIST$(-1.75) = 0.040059$. Do not reject the null hypothesis.
13. $T = -1.75$, p-value $=$ TDIST$(1.75, 19, 1) = 0.048126$. Do not reject the null hypothesis
14. 62.8% to 72.6%.
15. See Fig. 25.

	A	B	C	D	E	F	G	H
	Microsoft Excel - Book1							
	File Edit View Insert Format Tools Data Window Help							
	B7		=					
1	32.85234	CHIINV(0.025,19)						
2	8.906514	CHIINV(0.975,19)						
3	1.995292	19*3.45/A1						
4	7.359782	19*3.45/A2						
5	1.412548	The lower limit of the 95% confidence interval is given by =SQRT(A3)						
6	2.712892	The upper limit of the 95% confidence interval is given by =SQRT(A4)						
7								
8								

Fig. 25.

16. 0.0228
17. 0.0037
18. 0.0042
19. 0.0656

Chapter 1

1. (a) \$223.60 (b) \$5.0530 (c) 4.67 (d) 0.000001508. Reject the null hypothesis and conclude that the insurance has increased over \$200 for the period from 1995 to 2004.

2. Traditional: mean $= 81.13$ variance $= 54.8$ t-value $= -0.75$
   Experimental: mean $= 83.07$ variance $= 45.07$ p-value $= 0.23$
   Equal variances looks reasonable.
   Cannot conclude that there is any difference in achievement in the two groups.

3. (a) 4.10 (b) 5.53 (c) 2.35
   (d) p-value $= 0.044$. Do not reject at $\alpha = 0.01$.

4. (a) 0.009545 and 0.017182 (b) -0.0076364 (c) $z = -4.93$
   (d) p-value $= 0.000$. Conclude that the proportion in the aspirin group have fewer heart attacks.

5. (a) Company A: standard deviation $= 0.258996$. Company B: standard deviation $= 0.309966$.
   (b) $F = 0.698$.
   (c) p-value $= 0.441$. We cannot conclude that the population variances are different.

6. $Z = -3.344$ $= 2 * \text{NORMSDIST}(-3.344) = 0.000826$. Reject the null hypothesis and conclude that Diet 2 produces a higher mean weight gain.

7. $t = -0.74779$ $= \text{TDIST}(0.74779, 8, 2) = 0.47599$. Do not reject the null hypothesis.

8. $t = 14.6$ p-value is almost zero. Reject the null hypothesis.

9. $Z = 3.03$ p-value $= 0.002$.

10. $F = S_1^2/S_2^2 = 1.324$ p-value $= 0.496$ Cannot conclude that the population variances are unequal.

Chapter 2

1. Analysis of Variance for Time

Source	DF	SS	MS	F	P
Form	3	13.589	4.530	16.95	0.000
Error	36	9.622	0.267		
Total	39	23.211			

The p-value tells us to reject the hypothesis of equal means at alpha $= 0.05$. Figure 26 shows the relevant dotplots and boxplots.

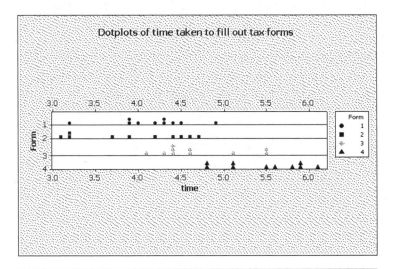

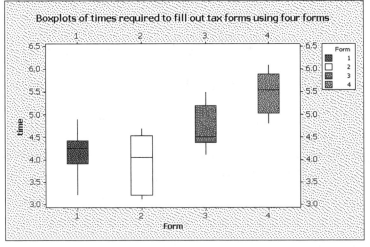

Fig. 26.

2. Tukey 95% Simultaneous Confidence Intervals
   All Pairwise Comparisons among Levels of Form

   Individual confidence level = 98.93%

   Form = 1 subtracted from:

   ```
   Form   Lower     Center    Upper    - - + - - - + - - - + - - - +-
   2      −0.8329   −0.2100   0.4129          (- - * - -)
   3      −0.0929    0.5300   1.1529              (- - * - -)
   4       0.6771    1.3000   1.9229                  (- - * - -)
                                       - - + - - - + - - - + - - - +-
                                       −1.2      0.0      1.2      2.4
   ```

   Form = 2 subtracted from:

   ```
   Form   Lower    Center    Upper    - - + - - - + - - - + - - - +-
   3      0.1171   0.7400    1.3629            (- - * - -)
   4      0.8871   1.5100    2.1329                 (- - * - -)
                                      - - + - - - + - - - + - - - +-
                                      −1.2      0.0      1.2      2.4
   ```

   Form = 3 subtracted from:

   ```
   Form   Lower    Center    Upper    - - + - - - + - - - + - - - +-
   4      0.1471   0.7700    1.3929             (- - * - -)
                                      - - + - - - + - - - + - - - +-
                                      −1.2      0.0      1.2      2.4
   ```

   Summarizing, we have: $\mu_1 = \mu_2$, $\mu_1 = \mu_3$, $\mu_1 < \mu_4$, $\mu_2 < \mu_3$, $\mu_2 < \mu_4$, and $\mu_3 < \mu_4$. Note 4 = Current. Or,

Form	2	1	3	Current

3. ANOVA: Time versus Block, Form

Factor	Type	Levels	Values
Block	fixed	4	1, 2, 3, 4
Form	fixed	4	1, 2, 3, 4

   Analysis of Variance for Time

Source	DF	SS	MS	F	P
Block	3	4.1669	1.3890	124.23	0.000
Form	3	2.7319	0.9106	81.45	0.000
Error	9	0.1006	0.0112		
Total	15	6.9994			

   $S = 0.105738$ R-Sq $= 98.56\%$ R-Sq(adj) $= 97.60\%$

```
Means

Block          N          Time
1              4          5.9750
2              4          5.4750
3              4          4.9250
4              4          4.6500

Form           N          Time
1              4          4.7500
2              4          5.0000
3              4          5.4500
4              4          5.8250
```

The *p*-values indicate that there are differences in the mean time required to fill out the tax forms for blocks and for form design. I would recommend form 1. The mean for that form is 4.75 hours. The current form requires 5.8 hours on the average.

4. **ANOVA: GPA versus Internet, TV**

```
Factor         Type       Levels   Values
Internet       fixed      2        0, 1
TV             fixed      2        0, 1
```

Analysis of Variance for GPA

Source	DF	SS	MS	F	P
Internet	1	0.7008	0.7008	14.02	0.006
TV	1	3.5208	3.5208	70.42	0.000
Internet*TV	1	0.0075	0.0075	0.15	0.709
Error	8	0.4000	0.0500		
Total	11	4.6292			

$S = 0.223607$ R-Sq = 91.36%   R-Sq(adj) = 88.12%

```
Means
Internet       N          GPA
0              6          3.2333
1              6          2.7500

TV             N          GPA
0              6          3.5333
1              6          2.4500
```

As can be seen from the interaction and main effects plots (Figs. 27 and 28), there is no interaction. High levels of either factor, Internet or TV, is detrimental to GPA.

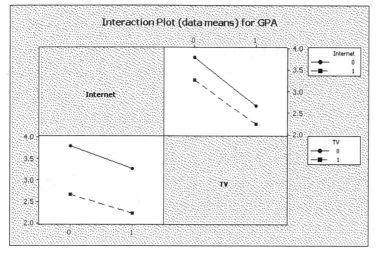

Fig. 27.

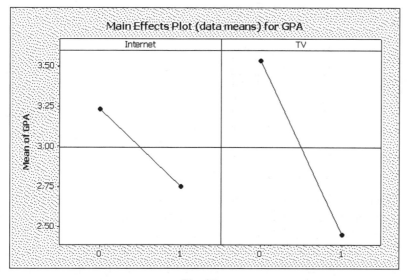

Fig. 28.

5. **ANOVA: Rating versus A, B, C**
Factor A is cheese, Factor B is meat, and Factor C is crust. (0 is low level and 1 is high level)

```
Factor     Type        Levels      Values
A          fixed       2           0, 1
B          fixed       2           0, 1
C          fixed       2           0, 1
```

Analysis of Variance for Rating

Source	DF	SS	MS	F	P
A	1	0.3306	0.3306	2.01	0.194
B	1	26.2656	26.2656	159.79	0.000
C	1	0.5256	0.5256	3.20	0.112
A*B	1	0.1406	0.1406	0.86	0.382
A*C	1	0.0056	0.0056	0.03	0.858
B*C	1	0.0156	0.0156	0.10	0.766
A*B*C	1	0.0056	0.0056	0.03	0.858
Error	8	1.3150	0.1644		
Total	15	28.6044			

$S = 0.405432$ R-Sq = 95.40%   R-Sq(adj) = 91.38%

Means

A	N	Rating
0	8	7.3750
1	8	7.6625
B	N	Rating
0	8	6.2375
1	8	8.8000
C	N	Rating
0	8	7.3375
1	8	7.7000

The parallel lines in the interaction plots (Fig. 29) confirms that there is no interaction. This is what the non-significant F values for $A * B$, $A * C$, $B * C$, and $A * B * C$ are telling us.

The main effects plot (Fig. 30) shows that meat is the factor that has the greatest effect. The interaction and the main effects taken together tell us that as long as meat is at the high level the ratings will be high, regardless of what the levels of factors A and C are.

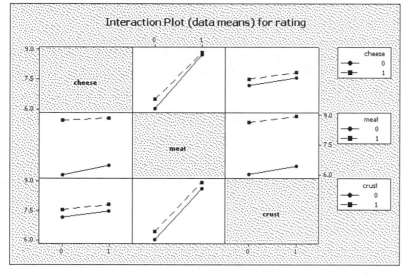

Fig. 29.

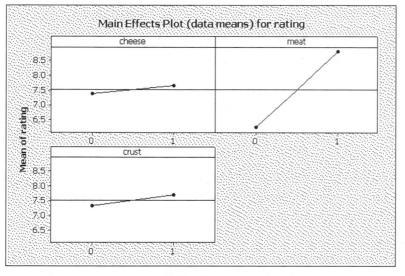

Fig. 30.

6.

Source of variation	Degrees of freedom	Sum of squares	Mean squares	*F*-statistic	*p*-value
Treatments	2	1156	578	2.28	0.133
Error	17	4304	253.1764		
Total	19	5460			

7.

Source of variation	Degrees of freedom	Sum of squares	Mean squares	*F*-statistic	*p*-value
Treatments	5	750	150	3	0.127
Blocks	5	500	100	2	0.233
Error	5	250	50		
Total	15	1500			

8.

Source of variation	Degrees of freedom	Sum of squares	Mean squares	*F*-statistic	*p*-value
A	1	50	50	2.29	0.1497
B	1	25	25	1.14	0.3015
AB	1	75	75	3.43	0.0826
Error	16	350	21.875		
Total	19	500			

9.

Source of variation	Degrees of freedom	Sum of squares	Mean squares	F-statistic	p-value
A	1	50	50	2.5	0.133
B	1	150	150	7.5	0.015
C	1	300	300	15	0.001
AB	1	15	15	0.75	0.399
AC	1	25	25	1.25	0.28
BC	1	20	20	1	0.332
ABC	1	5	5	0.25	0.624
Error	16	320	20		
Total	23	885			

10. The following pairs of treatment means are different:
1 and 4   1 and 5   1 and 6   2 and 4   2 and 5
2 and 6   3 and 6   4 and 6

The following pairs are not different:

1 and 2   1 and 3   2 and 3   3 and 4   3 and 5
4 and 5   5 and 6
Trt   1   2   3   4   5   6
_____
_____
_____

Chapter 3

1. (a) $y = 3.5x - 2$ (b) $y = -2x + 1$ (c) $y = 1.7x + 3.1$

2. (a) Slope $= 3.5$, y-intercept $= -2$
   (b) Slope $= -2$, y-intercept $= +1$
   (c) Slope $= 1.7$, y-intercept $= 3.1$
3. $y = 16.0 + 2.74x$
4. $y = 4.18 - 0.42x$
5. (a) A linear relationship is appropriate.
   (b) Positive.
   (c) $b_0 = 132.3$, $b_1 = 1.53$. Since no values for $x = 0$ were observed, b_0 does not have an interpretation other than it is the y-intercept of the regression line. The interpretation of b_1 is that each unit increase of drink increases the blood pressure by 1.53 units.
6. The computed test statistic is $t = ((b_1 - c)/(\text{standard error of } b_1)) = (1.528/0.613) = 2.49$. The critical value for 8 degrees of freedom is $t = 1.860$. Therefore the research hypothesis is supported.
7. The correlation coefficient is $r = 0.661$. The coefficient of determination is $r^2 = 43.7\%$. 43.7% of the variation in blood pressures is explained by variations in alcohol consumption.
8. 95% confidence interval (151.03, 174.76)
   95% prediction interval (139.05, 186.75)
9. (a) $\hat{y} = 121.03 - 2.07x$
   (b) Prediction interval (36.63, 122.63), confidence interval (67.34, 91.92)
   (c) $R\text{-Sq(adj)} = 61.2\%$
   (d) -0.795
10. (a) $\hat{y} = -1.53 + 1.60x$
    (b) Prediction interval (8.119, 14.414), confidence interval (9.667, 12.866)
    (c) $R\text{-Sq(adj)} = 87.8\%$
    (d) 0.945

Chapter 4

1. (a) yield $= -2.41 + 2.62$ fertilizer $+ 1.70$ moisture.
   (b) The intercept has no meaningful interpretation since there are no yield readings for 0 added fertilizer and 0 added moisture.
   (c) When moisture is held constant, an increase of one unit in fertilizer will increase yield by 2.62 units.
   (d) When fertilizer is held constant, an increase of one unit in moisture will increase yield by 1.70 units.
   (e) 16.9 units.

2. (a) See Fig. 31.

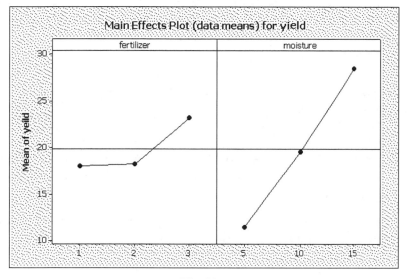

Fig. 31.

(b) See Fig. 32.

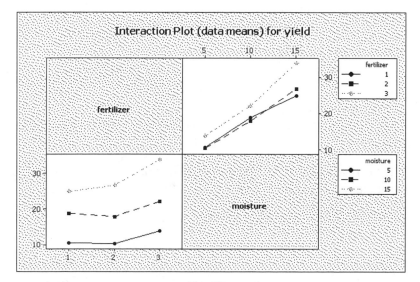

Fig. 32.

(c) Interaction cannot be tested for without replicating the experiment. Note that the total degrees of freedom $=8$, fertilizer degrees of freedom $=2$, moisture degrees of freedom $=2$ and interaction $=2 \times 2=4$, leaving 0 degree of freedom for Residual Error. Therefore, interaction model would be inappropriate.

3. (a) The regression equation is
   ```
   yield = −2.41 + 2.62 fertilizer + 1.70 moisture
   ```

Predictor	Coef	SE Coef	T	P
Constant	−2.411 2.329	−1.04	0.341	
fertiliz	2.6167 0.7913	3.31	0.016	
moisture	1.7033 0.1583	10.76	0.000	

   H_a: $\beta_1 > 0$ p-value $=0.016/2=0.008$ (the p-value for this test is a two-tailed p-value). Reject the null and conclude that fertilizer has a positive effect.

   H_a: $\beta_2 > 0$ p-value $=0.000/2=0.000$ (the p-value for this test is a two-tailed p-value). Reject the null and conclude that moisture has a positive effect.

   (b) Predicted Values for New Observations

New Obs	Fit	SE Fit	95.0% CI	95.0% PI
1	16.906	0.855	*(14.814, 18.997)*	(11.721, 22.090)

   Values of Predictors for New Observations

New Obs	fertiliz	moisture
1	2.50	7.50

4. The regression equation is
   ```
   Daily spam = −0.380 + 1.29 Year + 0.0500 Yearsq
   ```

Predictor	Coef	SE Coef	T	P
Constant	−0.3800	0.1918	−1.98	0.186
Year	1.2900	0.1462	8.82	0.013
Yearsq	0.05000	0.02390	2.09	0.172

   The years are coded 1, 2, 3, 4, and 5. Because yearsq is non-significant (p-value $=0.172$) the linear model is fit, with the following result.

   The regression equation is
   ```
   Daily spam = −0.730 + 1.59 Year
   ```

Predictor	Coef	SE Coef	T	P
Constant	−0.7300	0.1367	−5.34	0.013
Year	1.59000	0.04123	38.56	0.000

   The predicted number in 2004 is: daily spam $=-0.73+1.59(6)=8.8$.

5. Forward selection. Alpha-to-Enter: 0.05
   Response is Medcost on 6 predictors, with N = 30

step	1	2	3
Constant	−407.6	−186.7	−662.3
weight	13.4	13.5	18.0
T-Value	5.68	6.17	9.38
P-Value	0.000	0.000	0.000
exercise		−26.8	−47.9
T-Value		−2.37	−4.96
P-Value		0.025	0.000
smoker			−608
T-Value			−4.62
P-Value			0.000
S	402	372	281
R-Sq	53.57	61.59	78.90
R-Sq(adj)	**51.91**	**58.74**	**76.46**
C-p	26.1	19.1	1.7

6. The results with the interaction term will be presented first, followed by the model without the interaction term.

   With interaction:

   Regression Analysis: y versus x1, x2, x1x2
   The regression equation is
   y = −1681 + 0.946 x1 + 0.674 x2 − 0.000258 x1x2

Predictor	Coef	SE Coef	T	P
Constant	−1681.1	828.2	−2.03	0.089
x1	0.9460	0.3780	2.50	0.046
x2	0.6740	0.2707	2.49	0.047
x1x2	−0.0002575	0.0001197	−2.15	0.075

   S = 158.008   R-Sq = 71.7%   R-Sq(adj) = 57.6%

   Analysis of Variance

Source	DF	SS	MS	F	P
Regression	3	380262	126754	5.08	0.044
Residual Error	6	149800	24967		
Total	9	530063			

Without interaction:

Regression Analysis: y versus x1, x2
```
The regression equation is
y = 44 + 0.153 x1 + 0.106 x2

Predictor      Coef       SE Coef      T        P
Constant       44.0       255.6        0.17     0.868
x1             0.1533     0.1040       1.47     0.184
x2             0.10596    0.07378      1.44     0.194

S = 194.700    R-Sq = 49.9%    R-Sq(adj) = 35.6%

Analysis of Variance

Source          DF    SS        MS       F       P
Regression      2     264706    132353   3.49    0.089
Residual Error  7     265356    37908
Total           9     530063
```

The model with interaction is a better model.

7. *Model with interaction:*
```
Confidence interval(679.9, 1014.0)      width = 334.1
Prediction interval(425.8, 1268.1)      width = 842.3
```

Model without interaction:
```
Confidence interval(601.5, 947.5)       width = 346
Prediction interval(282.6, 1266.3)      width = 983.7
```

The confidence interval and the prediction interval are both narrower with the interaction term in the model.

8. $F = \dfrac{(SSE_R - SSE_C)/\# \text{ of } \beta s \text{ tested in } H_0}{MSE_C}$

$SSE_C = 1{,}998{,}721 \qquad MSE_C = 86{,}901 \qquad \# \text{ of } \beta s \text{ tested in } H_0 = 3$
$SSE_R = 2{,}057{,}532$

$(2057532 - 1998721)/3 = 19603.67$

$F = 19603.67/86901 = 0.226.$ p-value $= 0.877$
Do not reject the null hypothesis.

9. $F = \dfrac{(\text{SSE}_R - \text{SSE}_C)/\ \#\ \text{of}\ \beta\text{s tested in}\ H_0}{\text{MSE}_C}$

$\text{SSE}_C = 1{,}998{,}721 \qquad \text{MSE}_C = 86901 \qquad \#\ \text{of}\ \beta\text{s tested in}\ H_0 = 3$
$\text{SSE}_R = 7398969$

$(7398969 - 1998721)/3 = 1800082.67$

$F = 1800082.67/86901 = 20.714.$ p-value = practically zero.

10. The equation with interaction shows nothing significant at $\alpha = 0.05$.

```
The regression equation is
Y = -122 + 10.2 X1 + 4.30 X2 + 1.33 X3 + 0.120 X1X2 - 0.106
X1X3 - 0.0500 X2x3
```

Predictor	Coef	SE Coef	T	P
Constant	−121.80	60.27	−2.02	0.074
X1	10.160	5.735	1.77	0.110
X2	4.300	2.868	1.50	0.168
X3	1.3300	0.6789	1.96	0.082
X1X2	0.1200	0.1393	0.86	0.411
X1X3	−0.10600	0.06230	−1.70	0.123
X2x3	−0.05000	0.03115	−1.61	0.143

```
S = 6.96499    R-Sq = 78.8%    R-Sq(adj) = 64.7%
```

The model with no interaction shows moisture and fertilizer levels significant.

```
The regression equation is
Y = - 4.0 + 2.95 X1 + 0.950 X2 - 0.215 X3
```

Predictor	Coef	SE Coef	T	P
Constant	−3.97	17.18	−0.23	0.821
X1	2.9500	0.7842	3.76	0.003
X2	0.9500	0.3921	2.42	0.032
X3	−0.2150	0.1754	−1.23	0.244

```
S = 7.84246    R-Sq = 64.2%    R-Sq(adj) = 55.3%
```

Analysis of Variance

Source	DF	SS	MS	F	P
Regression	3	1323.70	441.23	7.17	0.005
Residual Error	12	738.05	61.50		
Total	15	2061.75			

The interaction plot is shown in Fig. 33.

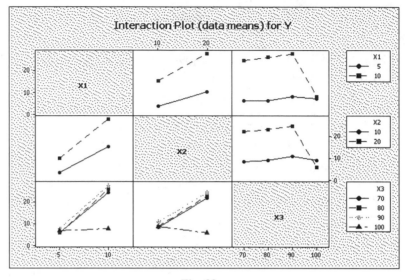

Fig. 33.

The main effects plot is shown in Fig. 34. When the temperature is 100 degrees, the yield is reduced for both levels of moisture and both levels of fertilizer.

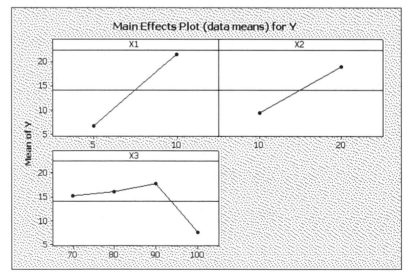

Fig. 34.

Chapter 5

1. Let $X =$ the number in the thirty trials greater than 30.5. p-value $= 2 * P(X \leq 5)$, where X has a binomial distribution based on 30 trials and 50% success probability. p-value $= 2 * \text{BINOMDIST}(5,30,0.5,1) = 2(0.000162) = 0.000324$. Reject the null and accept the alternative.

2. Let X be the number who would vote for A. The p-value $= P(X \geq 275)$, where X has a binomial distribution based on 500 trials and 50% success probability. p-value $= 2 * (1 - \text{BINOMDIST}(274,500,0.5,1)) = 0.028$. Reject the null and predict Candidate A to be the winner.

3. The median for the experimental group is 69 and for the traditional group it is 76. The p-value is 0.0001. The traditional group scored higher on the average than did the experimental group. Sum of ranks for the experimental group $= 429$, sum of ranks for the traditional group $= 846$.

4. **Wilcoxon Signed Rank Test: D**

	N	N for Test	Statistic	P	Median
D	30	26	228.5	0.182	0.5000

$D = \text{Husband} - \text{Wife}$
$T^- = 122.5 \qquad T^+ = 228.5$

5. Kruskal-Wallis Test on wtloss

Diet	N	Median	Ave Rank	Z
1	20	10.000	44.0	0.78
2	20	9.000	34.9	−1.25
3	20	9.000	40.3	−0.04
4	20	10.000	42.8	0.52
Overall	80		40.5	

$H = 1.83 \qquad DF = 3 \qquad P = 0.609$
$H = 1.85 \qquad DF = 3 \qquad P = 0.604$ (adjusted for ties)

There is no difference in weight loss for the four diets.

6. **Friedman Test: temp versus treatment blocked by block**

$S = 8.40 \qquad DF = 2 \qquad P = 0.015$

treatment	N	Est Median	Sum of Ranks
1	5	95.200	15.0
2	5	94.500	9.0
3	5	94.100	6.0

Grand median $= 94.600$.

The p-value is less than 0.05, indicating that the presence of the dog makes a difference in finger temperature.

7. Pearson correlation of nutritionists and housewives $= 0.467$. p-value $= 0.174$. Cannot conclude that there is a positive correlation between the two rankings.

Chapter 6

1. The test statistic equals 13.2778. The p-value equals 0.009995. The distribution differs from past years.

2. H_0: $p_1 = 1/3$, $p_2 = 1/3$, $p_3 = 1/3$. H_a: At least one of the p_i is not $1/3$.
   Test statistic $= 0.83333 + 3.33333 + 7.5 = 11.6667$.
   p-value $= 0.002928$. Reject the hypothesis that the die is balanced.

3. Chi-Sq $= 3.340$, DF $= 4$, p-Value $= 0.503$. Do not reject the null hypothesis.

4. Chi-Sq $= 2.744$, DF $= 6$, p-Value $= 0.840$. Do not reject the null hypothesis.

5. Chi-Sq $= 9.7385$, DF $= 7$, p-value $= 0.2039$. Do not reject the null hypothesis.

6. Chi-Sq $= 124.76$, DF $= 6$, p-value $= 0.000$. Reject the null hypothesis.

7. Chi-Sq $= 2.800$, DF $= 4$, p-value $= 0.592$. Do not reject the null hypothesis.

Bibliography

Keller, Gerald, *Applied Statistics with Microsoft Excel*, Duxbury, Pacific Grove, Calif., 2001.

Levine, David M., Patricia P. Ramsey, and Robert K. Smidt, *Applied Statistics for Engineers and Scientists Using Microsoft Excel and MINITAB*, Prentice Hall, Upper Saddle River, N.J., 2001.

Sincich, Terry, David M. Levine, and David Stephan, *Practical Statistics by Example Using Microsoft Excel and Minitab*, Prentice Hall, Upper Saddle River, N.J., 2002.

Spiegel, Murray R., and Larry J. Stephens, *Schaum's Outline of Statistics*, 3d ed., McGraw-Hill, New York, 1999.

Stephens, Larry J., *Schaum's Outline of Beginning Statistics*, McGraw-Hill, New York, 1998.

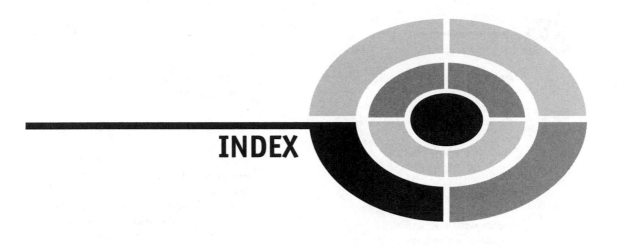

INDEX